NATTO

History, Science and Technology

Keith Wong

CHEMCAT
Toronto

CHEMCAT

Chemcat Publishing
Toronto, Ontario, Canada

Historical figures and photographs presented in Chapter I were published in Japan prior to 1957 and belong to the public domain under Japanese copyright law.
These include: Figure I.2, Figure I.3, Figure I.4, Figure I.5, Figure I.6, Figure I.7, Figure I.8, Photo I.1, Photo I.2, and Photo I.3.

All other photos, figures, and cover by Keith Wong.

ISBN-13 978-1-7389162-0-7
Hardcover
Version date: April 2nd, 2023.

To my family,

FYDP,

and my roommates, past & present

Contents

Introduction

Natto is made by growing the bacterium *Bacillus subtilis* variant *natto* on cooked soybeans. This process transforms the beans into a sticky, fragrant mass that can stretch to over five meters in length when stirred, giving its more specific name *itohiki natto* (糸引き納豆), or 'thread-pulling' natto. Invented and popularized as a breakfast food in Japan, it has a unique combination of desirable qualities and is simultaneously cheap, sustainable, convenient, and nutritious.

Consider this: in Japanese supermarkets a 50g pack of natto costs around 33 yen ($0.25 US Dollars)[a], and yet it has more protein than a large egg, without the cholesterol. Natto inherits all the nutritional benefits of soybeans, including its balanced amino acid profile, high fibre and mineral content, and essential fatty acids. It does this while simultaneously getting rid of the soybean's undesirable characteristics, including many of its antinutritional factors and its unpalatability, and adding some unique functional components and enzymes.

Why don't more people eat natto? Well, the answer seems obvious: on first encounter, even the most adventurous eaters examining its slimy, snot-like texture and peculiar coffee-cheese odour would hesitate to take a bite. Natto has a bit of a divisive reputation, considered an acquired taste on one hand, or utterly appalling on the other. Many people's first exposure to natto is probably through a video clip or newsreel, with subjects reacting to the gooey beans as they down a spoonful of the stretchy substance with a grimace.

If your first attempt at natto is its unseasoned, frozen-then-thawed form, you're going to have a bad time, and I wouldn't blame you for avoiding another encounter with the slimy beans. It's like having straight, powdered, imitation parmesan as your first cheese, and thinking that all cheese is awful.

Of course, there is a better way to eat natto. Most natto packages come with a salty, umami and sweet sauce (*tare*) that compliments its flavour and loosens the sticky texture. Users should eat natto with a warm bowl of rice, and add acidic or crispy toppings like green onions, kimchi, or nori, among many other possibilities. There is also an entire world of natto products that cater to individual tastes, including different sizes, colours and textures of soybeans, strains of natto bacterium that yield low odour or low stickiness, and a huge diversity of sauce options. Unfortunately, very few of these options are available outside of Japan, leaving only a handful of bottom-of-the-barrel imports.

There and Back Again: Natto's Import and Export Tale

Imported Japanese natto is also expensive. At most oriental supermarkets overseas, natto is relegated to the frozen section, where 150g (3x 50g packs) are sold for between \$3–\$5 USD, more than quadruple its cost in Japan of around 100 yen (\$0.75 USD). The freezing process also leaves it with an unappealing mealy texture when thawed, reminiscent of an old, bruised apple.

\$3 for 150g may not seem like much, but it's an absurdly high markup for an agricultural commodity that trades at around \$14 USD per bushel (27.2kg)[b]. Ironically, most of the mass market natto produced in Japan and destined for export, including those produced by large companies like *Mizkan, Shirakiku* and *Azuma*, rely on soybeans from the United States and Canada.

I live in Southern Ontario, Canada, and thanks to the relatively high moisture of air masses around the Great Lakes, the region is a major producer of soybeans, including small soybeans that are specifically grown for export to Japanese natto producers. When I buy a packet of frozen natto from the grocery store, the beans have likely travelled a minimum of 20,000 kilometres, or roughly half the earth's circumference, before reaching my bowl.

Problem Definition

This leaves the obvious solution: just make natto yourself. For any home cook that has a bit of experience with fermentation, this is simple enough, and there are some English books and online resources on the matter. However, most people don't want to deal with home fermentation. Natto is supposed to be a convenience food eaten within seconds of removing it from the fridge, just like yogurt.

Can natto be produced on a commercial scale, domestically? This has already been successfully done by multiple small producers. Still, without the economy of scale, automated machinery, and now over 100 years of industrial operations experience, natto produced and sold by western companies ends up costing even more than frozen, imported natto from Japan, even if using locally grown soybeans. Without a high demand, there won't be an economic justification for a large natto factory, and the shelf price of natto will remain high.

I'm still hopeful that a major natto factory will eventually be built in the west. The first large scale producers of *shoyu* and *miso* in the United States (*Kikkoman* in Walworth, Wisconsin and *Marukome*, in Irvine, California; respectively) were built and operated under the supervision of Japanese engineers once there was demonstrable market demand. Similarly, the first large throughput natto factory outside of Japan was built in Korea by *Pulmuone*, a company specializing in soy based foods, taking advantage of natto's increasing popularity in Korea in the mid-2000s. If a major natto producer were to set up a factory in the west, it might follow a similar fashion.

Understanding Natto

Japanese producers have perfected natto production and collected a wealth of knowledge around the underlying science. However, only a portion of this knowledge is available outside of Japanese language texts, and many of these sources were light on details. Most resources had no citations, or pointed to the same circle of websites with paper trails leading to dead ends. Preliminary research left me with lots of questions, such as:

1. Where did natto come from?
2. How is natto produced on industrial scales?
3. What makes good natto, and how is it eaten?
4. What's actually going on with the biology and chemistry during natto-making?
5. How valid are natto's health claims?

The objective of this book is to disseminate as much of this knowledge as possible, with a sufficient level of detail that can be used by producers, consumers, and researchers alike, to support an international natto industry.

Contents of this Book

This book is a literature review on natto and contains a full glossary and list of citations at the end. It relies on a combination of English and Japanese books and academic papers, some of which were fully translated thanks to powerful modern translation software. The chapters of this book are as follows.

Chapter I examines the history and origins of natto within Japan, starting at the dawn of agriculture and progressing into the modern era, when scientists successfully isolated the bacterium responsible for natto, developing an industrial process for mass producing the product.

Chapter II focuses on natto production. This includes details of each step of the modern natto making process, the equipment involved, the overall design of production facilities, as well as considerations for food safety.

Chapter III describes the finished package of natto. What makes good natto, and how is it usually eaten? We'll look at how to formally evaluate natto's quality, some of the marketing statistics, different sauces, and products made from natto.

Chapter IV focuses on the bacterium responsible for natto: *Bacillus subtilis*. This includes its cellular and molecular biology, and its relevance to natto production. It also looks at some practical questions like how the bacterium is isolated from nature, and how it has been bred specifically to make natto with desirable characteristics. We'll also look at fundamental questions like why natto contains high amounts of Vitamin K2 and an enzyme that can potentially dissolve blood clots called 'nattokinase'.

Chapter V looks at the soybean: *Glycine max*. This includes the soybean's seed structure, composition, and how some of its components, including its proteins and carbohydrates, are transformed into aromas and other functional components during the natto-making process.

Chapter VI details the nutritional aspects of natto. Producers of natto are quick to spew out a whole host of health benefits, but how many of these have been rigorously tested with sufficient evidence? We'll examine the basic nutritional characteristics of natto, and some of the research done on its unique physiological properties.

Finally, in Chapter VII, we'll look at natto-like products on a global scale. Japan is not the only region to apply the concept of 'bacteria grown on soybeans', and we'll look at other products from neighbouring regions, eventually coming back to theories on the origin of natto.

Further Reading

Science is constantly evolving, and the lexical difficulty of the topics covered in this book vary. For those looking for more fundamentals, or more details, I'll provide further reading from key sources and textbooks at the end of each chapter.

I hope that after reading this book, you'll gain an understanding of the history of natto, the minutia of science behind it, and the technology used for its production. Perhaps one day, I can walk into a grocery store and pick up a high-quality, 150g package of natto for $1.00, right between the tofu and yogurt.

---

[a] Exchange rate as of December 2022

[b] Commodity prices as of December 2022. Note that natto grade beans are more expensive.

I. History & Origins of Natto

In folklore, the story goes that the samurai Minamoto No Yoshiie (later known as Hachimantaro) discovered the stringy beans during the *Gosannen War* (ca. 1083CE) when he had set up camp for the evening and started cooking soybeans to eat. He and his men were suddenly ambushed, and they quickly packed up the soybeans into rice straw bags, slung them over their horses, and escaped into the night. The next day, the warmth of the horses and the rice straw transformed the boiled soybeans into natto.[1]

Most scholars find this origin story interesting, but are quick to point out that there isn't any written record from the *Gosannen War* of these events taking place. Alternative theories to the origins of natto pin its development to the spread of Buddhism, which reached Japan as early as 538CE.[1] This theory is supported by natto's etymology, roughly meaning 'temple beans', and there's evidence that Buddhist monks partook in the fermentation of soybeans, even if they weren't the inventors of natto itself.

Bacteria-fermented soybeans may have been discovered as soon as humans had cooked soybeans, and stored them in a warm place in close contact with any plant-based materials, which are a plentiful source of *Bacillus* endospores. Every traditional fermented food, including bread, pickles, wine, beer, cheese, yogurt, and others, was developed in a similar fashion. There is no discrete date of discovery for these products, and these foods arose from a combination of gradual agricultural developments, and the action of ubiquitous microorganisms in the environment.

There are many 'natto-like' fermented soybean products in Korea, China, the Himalayas, and the rest of East and Southeast Asia. These will be revisited at the end of this book, and for now, we'll only consider the origins and history of natto within Japan. We'll start by following the birth of agriculture on the Japanese islands from the last ice age and work our way towards the modern supermarket era.

Figure I.1 Map of modern-day Japan in East Asia, with commonly hypothesized routes of migration and approximate coastline during the Last Glacial Maximum

1. Prehistoric Japan

Prehistoric Japan is broken down into three periods based on artifacts discovered from each of these eras. These periods are approximate, and prehistoric theories are always being revised with newly discovered evidence. Archaeologists depend on artifacts pottery, tools, buildings, and land development projects, to piece together the story of Japan before written records.

Pleistocene and the Jomon Period

Humans first migrated to the landmass now known as Japan over land bridges or sea routes during the Pleistocene (the most recent ice age), when lower sea levels gave them easy access to the peninsula, most likely through the Korean peninsula from the south, or through the Sakhalin islands from the north. When the Pleistocene ended, the glacial ice melt caused sea levels to rise, and the Japanese archipelago was separated from continental Asia.[2]

The era that follows is known as the Jomon Period (ca. 14500BCE–900BCE), named after the distinctive cord-marked pottery from this period. Its inhabitants were primarily hunter-gatherers, but they also carried out some early forms of plant agriculture. This included the domestication of beans including the soybean, and the limited cultivation of grains including rice.[3]

It's possible that soybeans were fermented during this period, but this is difficult to ascertain without written records since soybeans themselves easily decompose without leaving much of a trace. We only know that beans were cultivated during this period from the small markings of beans inside of pottery. The more important achievement of the Jomon Period is that the inhabitants were practiced in plant husbandry, and possibly primitive forms of agriculture such as slash-and-burn (swidden agriculture), in which parts of forest were cut down, burned, and planted for one to a few years before being overtaken by the forest.[3]

Yayoi Period

Increasing contact with the continent lead to the transition into the Yayoi Period (ca. 900BCE–300CE), named after the Yayoi neighbourhood in Tokyo, where several artifacts from this era were discovered. This period is characterized by a new migration of people from the Korean Peninsula, who would mix with the native Jomon people and eventually become the ancestors of modern Japan, speaking a language like modern day Japanese.

This migration brought paddy rice farming, a highly productive cultivation method that involves the flooding of level fields and transplanting young rice seedlings, which could survive with their roots submerged while competing weeds would suffocate. Forming these fields requires a degree of coordination, but the resulting agricultural surplus lead to a population boom. Dry upland agriculture, including continued slash-and-burn agriculture, continued to be practiced for other crops such as soybeans, adzuki beans, peas, various kinds of millet, and barley.[3]

Rice cultivation creates a huge amount of rice straw, which was used for everything including sandals, tatami mats, ropes, and storage containers for grains and charcoal. Importantly, dried rice straw is a plentiful source of bacterial endospores, including the plant- and soil-dwelling bacterium *Bacillus subtilis.*

With the widespread use of rice straw as a general purpose packaging material, and plenty of soybean farming, a popular theory is that natto was discovered independently during the Yayoi period when cooked soybeans were left with rice straw. However, if this is the case, there is no evidence of the habitual production of natto as a food product.

Kofun Period

Prehistorical Japan concludes with the Kofun period (ca. 300CE–538CE), named after the keyhole shaped mounds built for members of the ruling class during this era. This period marks the maturation of a tiered society in Japan, and for the first time there is a marked difference between the burials of the ruling class compared to the commoners.[2,3]

By this time, the people in Japan are using fermented foods such as *sake* and vinegar, as well as *kuki* and *hishio*.[1] *Kuki* is the Japanese pronunciation for the Chinese character *shi* (豉), which is the oldest fermented soy food and equivalent to modern day Chinese *douchi.* This product is made from cooking, fermenting, then brining whole soybeans. *Hishio* is similar but made as a salted paste rather than keeping whole beans, and is equivalent to modern day *miso*. Both products require the intentional fermentation of cooked soybeans and other grains, so it's very likely that a product like natto was eventually developed as an offshoot of these fermented soybean products.

2. Classical and Feudal Japan

Japanese history begins around 538CE with the introduction of Buddhism from the kingdom of *Baekje*, which was one of three kingdoms located in modern day Korea (the others being *Silla* and *Goguryeo*). This brought with it the Chinese writing system, which eventually leads to the first documents of written history in Japan.[2]

It's during this period that we finally have written evidence of the existence of natto, along with the etymology of the word 'natto' itself.

Buddhist temples were also added to each feudal domain in Japan, and with a culture of vegetarian cuisine, there's strong evidence that the development of natto occurred in these temples.

Etymology of 'Natto': Temple Beans

The word natto (納豆) first appears in the book *Shin sarugaki ki* written by Fujiwara Akihara around 1058–1068CE. However, it was used in the term *shiokara natto* (塩辛納豆), or salt cured natto, which is equivalent to the abovementioned black fermented soybeans formerly referred to as *kuki*, indicating an etymological shift from the word *kuki* to *shiokara natto*.

The prevailing theory on the origins of the word 'natto' is that *na-* (納), stems from the word *nassho* (納所), which can refer to the Buddhist temple's storage area, the temple's kitchen, or the monk in charge of this area; whereas *-to* (豆) simply means beans. Thus, natto would roughly translate to 'temple beans' and would later evolve into two broader groups of terms, *itohiki natto*, referring to the non-salted threaded beans that are the subject of this book, and *shiokara natto*, referring to the brined beans used as a flavouring.

Legendary Folklore: Minamoto no Yoshiie

Around the same time ca. 1083, we have the natto legend of the samurai Minamoto no Yoshiie, who fought during the *Gosannen War*, also known as the 'Later Three Years War', although this is misleading since the war takes place between 1083–1087.[4] The war arises from a long standing quarrel between branches of the Kiyowara house in the northern Mutsu province, and Minamoto no Yoshiie was sent to quell them.

To reiterate the legendary origin story of natto, Minamoto no Yoshiie was planning to attack the enemy and set up camp in the evening. He and his men were cooking soybeans, suddenly ambushed by the enemy, and quickly packed up the cooked beans in rice straw bags before escaping. The very next day, he opened the bags of soybeans and was greeted by a unique aroma. He ate them and found them to be delicious.

Interestingly, when examining the Korean fermented soybean product *cheonggukjang*, which is almost identical to Japanese natto, there is a similar origin theory that these fermented soybeans were invented as a wartime food, since they can be fermented quickly.[5]

Early Records of 'Itohiki' and Food Fiction

It wasn't until much later in 1405 that the earliest surviving written record of 'thread-pulling' natto is found. The reference was in *Noritoki kyo ki* ('Diary of Noritoki'), which belonged to a nobleman. The diary entry on December 19, 1405, mentions *itohiki daizu* (string-pulling soybeans). This refers to the same product as *itohiki natto*, since only biofilm forming bacteria will create sticky threads when fermented. This document is the earliest surviving reference to the term *itohiki* in reference to soybeans.[1]

If you find this nobleman's diary boring, another intact record of *itohiki natto* from this time period is a fictional piece called *Shojin Gyorui Monogatari* ('Tale of vegetables and fishes' ca. 1450), in which anthropomorphic vegetarian and animal foods get together and have an epic battle.[6]

The beginning of the story is as follows: a formal ceremony is taking place at the imperial palace, in the first year of the fictional 'Fish-Bird' era. The Roe brothers, sons of Lord 'Osuke' Salmon Long-Fin, arrive at the ceremony and are enraged when they see that Natto 'Taro' Big-Seed, son of Lord Soybean, is seated closest to Shogun Rice (the de-facto monarch). Meanwhile, the Roe brothers have been demoted to the lowest seats, and they decide to return home in disgust.

The Roe brothers' father, Lord 'Osuke' Salmon Long-Fin, can't tolerate this humiliation and assembles an army of fishes, mollusks, birds, and animals to attack Lord Soybean, whose own army consists of seaweeds, vegetables, roots, beans, sweets, and fruits.

No one is sure of the story's intended audience or purpose, though most scholars agree that *Shojin Gyorui Monogatari* is a satire or parody, possibly of *Heike Monogatari* compiled around a century prior ca. 1330. It could be about the stupidity of war and politics, or an argument between the conflicting ideologies of Buddhist vegetarianism *(Shojin ryori)* and carnism or omivorism.

Nevertheless, between the writings of a nobleman's diary, and a satirical fiction depicting natto as a main character, we know that by the 1400s natto had become a mainstay food in Japan.

***Sengoku* Period and European Contact**

Perhaps *Shojin Gyorui Monogatari* was more prophetic than it intended, because the era that immediately followed was the *Sengoku* period (1467–1615), in which a power vacuum following the *Onin War* in 1464 left Japan in a period of civil war with local clans vying for power.[2]

During the *Sengoku* period, the Portuguese make contact with Japan in 1542, becoming the first Europeans to reach Japan and starting the *Nanban* trade, which brought about a cultural exchange that included Christianity, firearms, and European shipbuilding technology.[2]

This period would last until the isolationist policies of *Sakoku* during the *Edo* period. By 1603, near the end of the *Sengoku* period, the Portuguese had compiled a Japanese-Portuguese dictionary called *Vocabulario da lingoa de Iapam*, in which Natto is defined as follows:[1]

Nattô: type of food made by a brief boiling of grains / seeds (note: soybeans were used), which are then put into an incubation chamber (*muro*).

This is the first European written record of Japanese natto.

3. Edo Period

The Edo period was ushered in by the *Tokugawa Shogunate* under Tokugawa Ieyasu, who consolidated power and stamped out social unrest, sometimes brutally, bringing an end to the *Sengoku* era of civil war and ushering in a period of relative peace and stability. Isolationist policies known as *Sakoku* were also implemented, effectively cutting off Japan from the rest of the world, and foreigners could only trade though a handful of selected channels.[2]

During the first century of the Edo period, the population of Japan doubled to nearly 30 million. City populations grew, as did civil engineering projects, increased literacy, and flourishing culture.[2]

With increased literacy and writing, and without clans burning each others' castles to the ground, this era contains many intact records about natto with plenty of detail.

One important work is *Yorozu kikigaki hiden* ('Secrets of things heard and written') ca. 1650, in which 34 of its 54 sections relate to food preservation and preparation. It includes instructions on how to make natto, among other fermented foods like *shoyu* and *miso*. This handbook was incredibly popular with nine successive printings made between 1650–1673.[1]

We also know that natto was a common and inexpensive street food for Edo period vendors, selling for a fraction of the cost of tofu. The etymology of natto continues to develop with more detail. Several new terms are mentioned describing natto in its various forms, and distinguishing it clearly from the salted *shiokara natto.*

Natto would be peddled by vendors called *natto-uri*, and sold in the form of *tataki natto*, which was chopped up *itohiki natto* mixed with greens and tofu, or as *tsubu natto,* which is simply whole grain natto.[1]

Illustrations of Natto

Several important illustrations were compiled during the Edo period containing natto in the context of everyday life. One of them is found in *Jinrin kinmō zui* ('Illustrated dictionary of humanities and social sciences'), dated ca. 1690, which is a seven-volume compilation of descriptions and paintings depicting everyday life in Edo.[1]

One illustration shows two men, each carrying a shoulder pole, at the ends of which are hanging boxes or baskets of food. One man is described selling *horo miso,* and the other described selling *tataki natto*.

Most illustrations of natto from the Edo period are similar, and depict a walking vendor with a shoulder pole holding wooden boxes or baskets on each end, chanting '*natto... natto... natto...*', which would become an iconic representation of the food throughout this era.

Figure I.2 Left: seller of horo miso. Right: seller of tataki natto.
From Jinrin kinmō zui, ca. 1690 (early Edo period)

Production and Sales in the Edo Period

Based on the written and illustrated record, we have an accurate idea of how natto production and sales took place during this period. Soybeans would be soaked overnight then boiled for around 5 hours until soft, and then packed in bundles of straw known as *tsuto* and put in a cellar, in the middle of which a small fire would be lit.

The next morning, the natto would be sold in their rice straw bundles, or in a bamboo basket, from which portions of natto could be scooped out using a rice scoop *(shamoji)*. It would be sold on the streets of Edo and Kyoto by *natto-uri* who would chant, "*natto… natto…*" whilst carrying the natto on their shoulder poles. This natto, often sold chopped and mixed with other ingredients, would sell for one-fifth of the cost of tofu on a weight basis, making it a cheap street food and common breakfast item.[1]

Two ways to eat this natto were mixed into a soup called *natto jiru*, or eaten on top of rice with salt, *shoyu* and mustard serving as popular condiments at the time.

Natto was sold primarily during the winter and made by the peasant class in farmhouses. From a technical standpoint, this makes sense for a few reasons. Firstly, the process of boiling soybeans at atmospheric pressures is energy intensive, requiring around five hours of boiling compared to adzuki beans or green beans, which require less than an hour. In typical Edo-style farmhouses, this could take place on the central hearth called *irori*, providing heating to the home, so no energy would be wasted.

The cold weather would also help halt fermentation following the incubation process in the heated cellar. Without cooling, natto can quickly over-ferment in warm conditions, becoming bitter and taking on a strong ammonia odour.

Producing natto during the winter also helps to provide food during a time of year when fresh vegetables are not readily available, and gives the farmer an additional source of income and activity during the winter months. For similar reasons, *sake* brewing was also a winter activity. Later in the Edo period, natto would also be produced in the summer.[1]

Commentators note that natto making during the *Edo* period was prone to failure. Any ubiquitous microorganism had a chance of colonizing the cooked soybeans, often leading to undesirable or unpredictable results. This problem would remain until start of the 20th century, when modern science finally leads to sanitary food production and the development of pure natto inoculum.

4. Modern Japan

Matthew Calbraith Perry of the United States of America arrived in Edo Bay via the gunboat *Mississippi* on July 14, 1853, accompanied by three other ships, and armed with huge Paixhans shell guns and coal-fired engines powering a massive paddlewheel. His fleet steamed past Japanese lines, refused any requests to turn around or sail to Nagasaki (the designated trading port for foreigners), and requested to deliver a letter from U.S. President Millard Fillmore requesting an end to the isolationist policies of Japan. The Japanese were rightly frightened: not only because Mr. Perry threatened to turn his ships to Edo and burn it to the ground, but also because at that moment, the Japanese knew they were lagging far behind in technology.[2] Matthew Perry was able to deliver the letter, and promised to return the following year for a reply.

Before Matthew Perry's second visit, the Tokugawa shogunate learned about the Opium wars, and how China had been flooded with opium, and then forced to make territorial concessions when they tried to ban the substance and were completely outgunned. The shogunate realized that if they wanted to avoid a similar fate, they would have to make peace with the Europeans and Americans, and play catch up in technology. Mathew Perry would return as promised in 1854 and force the signing of unequal treaties at the Convention of Kanagawa. Thus ended the Edo period and isolationist policies of *Sakoku,* and began the new era of Modern Japan.[2]

There are entire books dedicated to the subsequent period known as the Meiji Restoration, but in short, during this time the entire political and social structure was overhauled, with power being restored to the Emperor. Japan rapidly industrialized and adopted western ideas, including modern western science. During this time many of Japan's Universities were founded, and in the subsequent years the first academic articles related to natto were written and published.

The speed at which Japan develops during and after the Meiji Restoration is astounding. Consider this: in a single lifetime, someone born in the Edo period can go from eating straw-packed natto on the streets of Edo, to eating refrigerated, polystyrene packaged natto which had beans cooked in a steel pressure cooker, inoculated from a pure culture of *Bacillus subtilis* endospores isolated in a lab.

The Modern Japanese era, spanning from 1868 to present day, starts with the Meiji Restoration (1868–1912), followed by the Taisho period (1912–1926) and Showa period (1926–1989), during which two World Wars were fought, followed by a long period of relative peace and a post war economic boom. This brings us to the modern Japanese periods of Heisei (1989–2019) and Reiwa (2019–present). It's during this time that rapid scientific advances were made, and natto evolves from a farmhouse product to an industrially produced supermarket commodity.

'Natto Thoughts' Prior to Industrialization

In the early years of the Meiji restoration (ca. 1880s; the publication date of the original book is not known), a book titled *Nattō kō* ('Natto Thoughts') was written. This is a book entirely dedicated to the topic of natto, and provides of a glimpse of the traditional natto industry right before its transition into the modern era. The author also illustrated several *natto-uri*, some of whom wore traditional clothes typical of the Edo period, while others were dressed in modern, western style clothing.

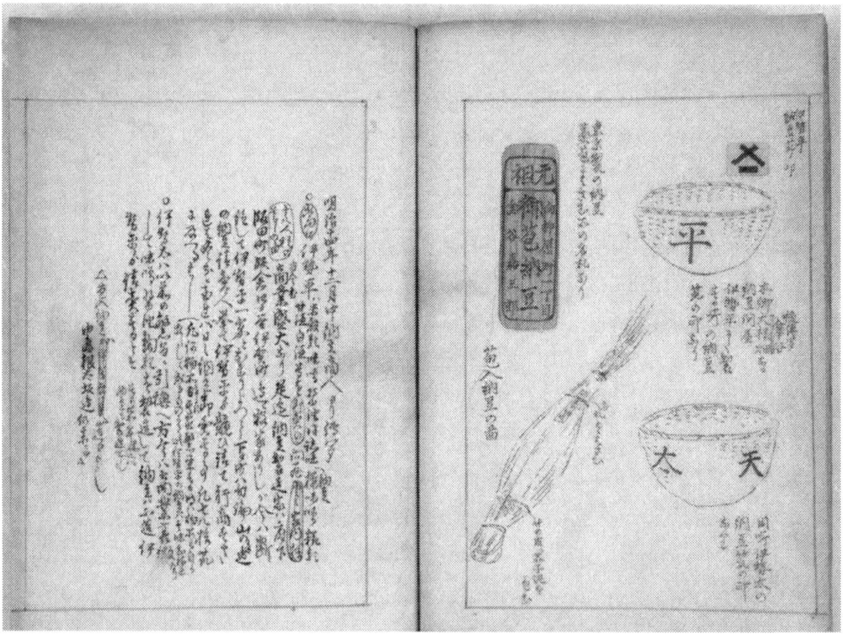

Figure I.3 Pages from 'Natto Thoughts' ca. 1880s
Left page: the text indicates the depictions are from mid-December, Meiji 12 (1879)
Right page, top left: a wooden or bamboo tag inserted into a straw package of natto.
It reads: 元祖 / 御苞納豆 / 向柳原町一丁目 / 金谷喜三郎 *(Maker / Straw natto / Mukōyanagiharachō-1-chōme [Address] / Kanaya Kisaburō [Company])*
Right page, bottom left: natto packaged in a rice straw bundle (tsuto).
Right page, right: bamboo baskets used to hold natto, from two different makers.
If not selling in rice straw bundles, natto-uri would scoop natto out of these baskets with a rice scoop (shamoji).

The natto itself hadn't changed much from Edo period offerings. It was still packed, fermented, and sold in rice straw bundles or baskets. The book contains several examples of package tags and labels from this period.

Figure I.4 Three depictions of natto-uri from 'Natto Thoughts'
Left: Natto-uri ca. Meiji 16 (1883), dressed in a western uniform.
Right, top: Natto-uri from 'Isehira' (伊勢平), dressed in traditional clothing.
The caption describes the vendor selling miso and shoyu moromi (mash) in addition to natto.
Right, bottom: Natto-uri from 'Kanaya' (金谷), dressed in western uniform.

4.1 Early Research into Natto

The first research papers on natto follow in the footsteps of several developments in microbiology occurring in Europe at the time. These include the work of Louis Pasteur, who between 1860–1864 disproved 'spontaneous generation': the idea that microorganisms would suddenly develop from non-living matter. He did this by demonstrating that a sterilized and sealed flask would not allow microorganisms to grow, but one that had access to air would. This in turn supported the germ theory of diseases, which is the idea that microorganisms were the cause of infectious disease. The competing theory at the time was that 'miasma' — or foul-smelling air — was the cause of disease. The germ theory was further developed by Robert Koch, who among a great many number of discoveries, ushered in the 'golden age' of bacteriology when he discovered *Bacillus anthrax* and its life cycle in 1875, with a paper published the following year.[7] *Bacillus subtilis* is also isolated and named during this period, the details of which are presented in Chapter IV.

Koch's other developments include greatly improving the techniques of microscopy, especially for bacteria, and developing plating methods with the help of Walther and Fanny Hesse, whom developed agar as a growth medium, and Julius Richard Petri, who developed the Petri dish.[7] These basic microbial techniques would pave the way for future microbial analysis.

On the Vegetable Cheese, Natto

In 1894, Kikuji Yabe of Tokyo University became the first researcher to complete a detailed scientific analysis of natto. Yabe describes the natto production method of the time and conducts a microbial and chemical analysis of the product. The paper was published in three journals, including an English article titled "*On the vegetable cheese, Natto*".[8]

The paper summarizes the natto-making process from the Edo period. To reiterate, soybeans are boiled for five hours before being portioned into straw bundles and placed in a cellar, in the middle of which a small fire is lit, and the cellar is closed for 24 hours.

Yabe sampled the sticky material of natto and plated it on a petri dish, identifying four different types of cultures which were then subcultured. He identified three micrococci, with yellow, yellow-orange, and white colonies, and one non-motile bacillus. He couldn't determine which of these were the primary cause of natto, but noted that one of the yellow micrococci seemed to cause the odour of natto when grown on soybeans.

On the chemistry side, Yabe noted that natto had a moisture content of 59.12%, and compared to raw soybeans, had a slightly higher (2%) total nitrogen content, while the soybean proteins were significantly transformed, with 42% of proteins being degraded into smaller peptones and peptides. Based on this information, he guessed that natto would have a higher digestibility than boiled soybeans.

Finding the Responsible Microorganism

Shin Sawamura of Tokyo University followed up the work of Yabe in 1906, publishing *"On the micro-organisms of natto"*.[9] Sawamura was determined to identify exactly which microorganism was responsible for natto. He collected samples of natto from several producers and plated them. Then, he took the individual colonies and propagated them on sterilized, cooked soybeans. Of the many organisms tested in this process, Sawamura identified two notable species of Bacillus.

Bacillus No. 1 was noted as the organism that produced the distinctive aroma of natto, but on its own, didn't create much of the viscous substance. This bacillus was motile, and had flagella which allowed it to swim. Bacillus No. 1 was present in great numbers in the samples he collected, so Sawamura concluded that this was the organism chiefly responsible for natto. He named the organism *Bacillus natto.*

Bacillus No. 2, on the other hand, produced plenty of the viscous substance but a less agreeable taste. He also noted that this bacillus produced an amylase (then referred to as a *diastatic* enzyme) that yielded sugars in the presence of starch.

Today's knowledge of bacteria would likely identify Bacillus No. 1 as a motile strain of *Bacillus subtilis*, while Bacillus No. 2 would be a non-motile *Bacillus subtilis* or the closely related *Bacillus amyloliquefaciens.*

Shunsuke Muramatsu of Tokyo University, building on the work of Sawamura, published a detailed English-language paper in 1912 titled "*On the preparation of natto*"[10] describing all aspects of natto known up to that time, and also proved that that distinct odour and viscidity of natto could be brought about by a single bacillus. By modern definitions, he had isolated a non-motile, poly-γ-glutamic acid positive version of *Bacillus subtilis* similar to modern natto- producing strains.

Sawamura later confirmed Muramatsu's findings in a 1913 paper titled "*On Bacillus natto*",[11] isolating a single bacterium that caused the odour and stickiness expected of natto. Sawamura would name the isolated strain of bacterium after himself, as *Bacillus natto* SAWAMURA.

4.2 Developing Modern Natto

The Father of Modern Natto

Jun Hanzawa of Hokkaido University is often credited as the father of modern natto production. Hanzawa had just returned from studying microbiology in Europe between 1911–1915, and returned to Hokkaido University to start a microbiology course at the Department of Agricultural Chemistry.

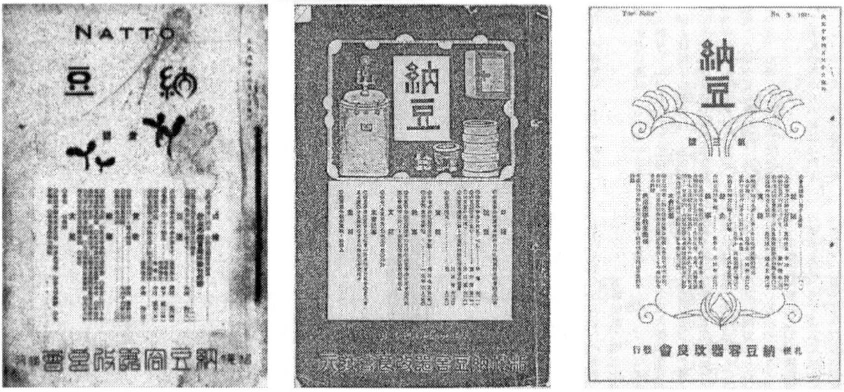

Figure I.5 Jun Hanzawa's magazine 'Natto'. Left to Right: 1st, 2nd, and 3rd articles.

Hanzawa questioned the use of rice straw, which was full of dust, contaminants, and many microorganisms, not just those required to make natto. He pointed out that natto made with straw sometimes ended up with a foul odour, which could be so offensive that even cats and dogs wouldn't dare touch it.

In the span of 1919–1921, he released three articles in his magazine simply titled, "*Natto*", which were published by an association he founded in 1918. This association was called the 'Sapporo natto container improvement association', which he used to advocate the abolishment of rice straw packaging in natto.[12–14]

Working simultaneously as a microbiologist, an extension worker for farmers, and a pilot plant operator, Hanzawa made important developments that would lead to the modern natto production process. Firstly, he developed a pure culture inoculum. Building on the work of previous researchers, the

Figure I.6 Label of Jun Hanzawa's Daigaku Natto (University Natto) ca. 1920

inoculum was made by simply scraping colonies of natto bacteria off an agar slant, and suspending it in sterilized water or the cooled broth from cooking beans.

Without the need for straw as a source of microorganisms, Hanzawa did away with rice straw as a packaging material. Instead, he developed the use of paper-thin sheets of pine veneer (*kyogi*) or small boxes of pine wood (*oribako*) as packaging.

Hanzawa had an interesting motivation for developing this natto process. To fund the operating costs of his microbiology projects, he sold this pure cultured natto on campus as *Daigaku natto* (University natto). He would use these profits to further develop the modern natto making method.

Natto Associations and Process Refinement

Hanzawa would leverage his 'Sapporo natto container improvement association' to publish a book titled 'Natto Manufacturing Method', which was first released in 1926. He would host workshops for natto producers, teaching them the new methods and convincing producers to abandon rice straw natto.[12–14]

Figure I.7 Label of Jiro Miura's Miyagino Natto ca. 1920

One of Hanzawa's disciples, Jiro Miura, founded Miyagino Natto in 1920–1921. Working closely with Hanzawa, he realized that temperature control in the fermentation room was difficult and prone to overheating from the metabolic heat of the bacteria. Miura designed an improved fermentation room with a top vent to help with cooling and dehumidification in the latter phases of fermentation.

By the 1930s, Miura would operate the most advanced natto facility of the time. Miyagino Natto still exists today and is one of the three major producers of commercial natto *Bacillus subtilis* endospores known as *natto-kin*. Latter editions of Hanzawa's book 'Natto Manufacturing Method',[15] published in 1930 and 1936, covered the fundamentals of natto production, the concepts of which are still relevant today. The main steps of the natto making process outlined in this book are as follows.

1. Sacks of dried soybeans would be brought into the facility.
2. A worker would visually examine the soybeans on a tray, manually sifting out any pebbles, debris, and discoloured or damaged soybeans.
3. Sorted soybeans would be thoroughly washed and soaked.
4. Soaked soybeans would be pressure-steamed in a direct-fired, steel pressure cooker fixed atop a brick stove.
5. Steamed soybeans were brought into a rotating drum to be inoculated with a sprayed solution of *natto-kin.*
6. Inoculated beans would be filled into individual boxes, with wooden *oribako* boxes being common. These small boxes would be loaded onto wooden trays similar to those used for *koji* known as *kojibuta.*
7. The trays of boxed, inoculated beans would be brought into an incubation room for fermentation.
8. Once fermentation was complete, the boxes of natto would be peddled out on the streets immediately that morning.

The book also contained details on temperature and humidity measurement, detailed work diaries, and design features of the natto factory.

Shunsuke Muramatsu and Kintaro Naruse would publish their own book titled 'Latest Natto Manufacturing Method' in 1937 and 1950.[16] The two would establish the Naruse Fermentation Chemistry Laboratory Co. Ltd. in 1946, which is another one of the three major producers of *natto-kin*.[17]

The remaining major *natto-kin* producer is Yuzo Takahashi Laboratory Co. Ltd. Unfortunately, the history of this company is unclear, other than their website stating that they were founded in 1935.[18] The company was founded by Mitsuji Takahashi, who worked as a natto producer selling a product called *Tanikaze natto* which was later named *Sendai natto*.

Working together, natto producers cooperated by forming associations, helping to standardize the process, and exchange ideas. In 1939, prefectural 'Natto associations' were formed, eventually consolidating into the 'National Natto Industry Association' in 1941 and then renamed the 'Japan Natto Cooperative Society Federation' in 1954. This organization remains today, with 103 member companies as of 2020.[19]

Figure I.8 Process steps from 'Natto Production Method', remastered, ca. 1926–1936

Photo I.1 Direct fired pressure steaming kettles ca. 1926–1936

Photo I.2 Labeling wooden oribako containers of natto ca. 1926–1936

Photo I.3 Natto peddled by motorbike, ca. pre-World War II
Side of bike: 'Modern Scientific Natto'. Note that right-to-left horizontal reading was common.

Natto in the Postwar Period

In the era following World War II, there was a recovery period followed by a postwar economic boom that brought in massive productivity increases and a transformation to the Japanese economy.

For the natto industry, the most significant changes to the production process were the consolidation of producers, the widespread use of refrigeration, and improvements to packaging and mechanization.

From the 1960s to the 1990s, the food industry in Japan would become consolidated, with the expansion of supermarkets requiring large volume bulk deliveries that could only be achieved by larger and larger factories. These factories also benefited from economy of scale, and used increasingly high throughput process equipment. An ongoing labour shortage and associated increase in wages meant many tasks needed to be automated for factories to stay competitive, and several machines were developed.[12–14]

Industrial equipment manufacturers played an important role during this period. Founded in 1959, one of these companies, Suzuyo Kogyo Co., Ltd., remains the dominant player in natto making equipment today.[20]

Rapid improvements in factory equipment took place during the 1960s, including the introduction of fibre reinforced plastic (FRP) incubation rooms in 1968 with automatic heating, ventilation, and air conditioning.[14]

By 1974, natto makers would adopt the use of polystyrene paper (PSP) packs, followed by paper cups in 1978. These package types, combined with high speed automatic filling machines introduced in 1981, created significant labour savings and replaced the wooden boxes and packages of the prewar period. Starting in 1986, the widespread use of factory automation was adopted, introducing control panels and Programmable Logic Controllers to the factory floor.[14]

For consumer convenience, most natto is also packed with small packets of sauce and mustard, or with sauce integrated with the package lid. Wooden packaging is still used with small-scale producers producing high quality natto to distinguish themselves as a 'craft' product.

Refrigeration led to increased consistency and quality of natto. Freshly incubated natto packages could be immediately cooled down, halting the fermentation process, and allowing the natto to cure at low temperatures, ridding it of any off-odours while maintaining its quality. Refrigeration in the supply chain meant that these natto products could be kept cold in the factory, truck, supermarket, and eventually to people's homes, which were seeing widespread adoption of the refrigerator between 1960–1990.[21]

This entire process, pioneered by Jun Hanzawa and optimized throughout the rest of the century, is described in Chapter II: Natto Production.

During this time, Japan would significantly shift away from agriculture, focusing increasingly on manufacturing and services. Before WWII, almost every single soybean in the natto process was grown in Japan, but nowadays most natto is made from imported soybeans from agricultural powerhouses such as the United States and Canada. The agricultural departments of these countries would develop varieties of soybeans bred specifically for natto, for direct export to Japan. Meanwhile, natto made using Japanese soybeans is specifically marketed as such, and fetches a higher price.

Academic research in natto continues to the present day. Many of the details involving the chemistry, biology, and health effects of natto have been studied in detail, following in the footsteps of scientific advancements after WWII. These are presented in Chapter IV–Chapter VI.

Summary

It's difficult to pinpoint the origins of natto in Japan. Rice and soybeans had already been cultivated on the Japanese archipelago by the Jomon period, followed by more sophisticated agricultural techniques in the Yayoi period, during which there would have been plentiful access to soybeans and rice straw.

The earliest written record of the word 'natto' dates to ca. 1050, and by the 1400s we are certain that natto had already become a commonplace food. natto was peddled in the mornings during the Edo period, and scientific research on natto began after the Meiji restoration, eventually developing into the Modern natto industry.

Further Reading

One of the most complete English language archives of Natto's history (and other soy foods) can be found on William Shurtleff and Akiko Aoyagi's compendium:

- **The History of Natto and its Relatives (1405–2012)**[1].

Meanwhile, the Japan Natto Cooperative Society Federation has published a modern history on natto:

- 納豆近代五十年史 [**50 Years of Modern Natto History**]

Finally, a Japanese natto enthusiast named Taiji Ishii (石井泰二), who goes by the online handle @taiji141 runs a fantastic blog called 'Shichiten Natto' (七転納豆) and wiki called natto wiki (納豆 wiki), containing plenty of details on the history of natto, as well as a massive compendium of natto products. His help has been greatly appreciated, and his blog and wiki contain photos and historical documents on natto beyond what was shown in this chapter. He has also published a book titled *Natto Kurabe* (納豆くらべ).

- **blog.livedoor.jp/taiji141/** (blog, 七転納豆)
- **seesaawiki.jp/w/taiji141/** (納豆 wiki)

II. Natto Production

The modern natto production process, which involves growing a pure culture of *Bacillus subtilis* on pressure steamed soybeans, was pioneered by Jun Hanzawa and Jiro Miura in the 1920s and has been gradually improved.

The objective of natto making is to allow *B. subtilis* to proliferate, developing the flavour and texture of natto, while ridding the soybean of its raw, beany quality. The fermentation process also needs to be stopped before it progresses too far. Note that in biochemistry, fermentation refers to processes occurring in the absence of oxygen. For the purposes of this book, we'll adapt the broader definition in which fermentation is defined as the activity of microorganisms bringing a desirable change to a food.

If natto is over-fermented, the soybean's soluble sugars, the bacteria's preferred energy source, become depleted. The bacteria will then increase its consumption of amino acids as its energy source. This process requires the deamination of amino acids, liberating ammonia, which has an irritating, pungent odour. Ammonia is always produced in some quantity during natto fermentation, but the exhaustion of soluble sugars forces the bacteria to rely on amino acids more for energy, producing excessive ammonia.

Unlike the hyphae of filamentous fungi used in Japanese *koji* (*Aspergillus spp.*) and Indonesian *tempeh* (*Rhizopus spp.*), the small cells of *B. subtilis* cannot burrow into the soybean's seed structure, and instead form a colony on its surface that develops into a biofilm. The bacteria rely on diffusion to transport its excreted enzymes into the soybean, while soluble carbohydrates and amino acids dissolve back into the growing biofilm. This biofilm consists mainly of two polymers: poly-γ-glutamic acid and levan, the former of which is responsible for natto's stretchy mucilage.

The selection of soybeans is an important factor affecting the quality of the product natto. The soybeans should be high in soluble carbohydrates to provide *B. subtilis* with a food source, whereas a greater surface area to volume ratio increases diffusion and speeds up the progression of bacterial growth.

The fermentation process only takes around 18–20 hours, depending on a few factors, and the whole process takes a total of four days, including the soaking of beans, and refrigerating the natto to mature after fermentation.

All natto factories use the same basic equipment, regardless of their throughput. The largest factories can process upwards of 15 tonnes of soybeans a day, yielding around 600,000 individual 50g packages. The ten largest natto companies in Japan account for 85% of sales, with heavy use of factory automation and an economy of scale that allows them to produce natto at around 100 yen for a three-pack.[20]

Natto can be produced at much smaller scales, at a higher price. To differentiate themselves, small natto makers will often focus on unique offerings like domestic (Japanese) soybeans, including specialty beans (for example green, red, and black heirloom soybeans), unique sauces and mix-ins, and sell locally, charging a premium for their product. At small production scales, there is less use of automated equipment, since this equipment can have high capital costs.

The natto making process is simple, so the challenge for producers is not to simply make natto, per se, but to do so consistently and economically without contamination or mistakes. Customers purchasing the same natto label every grocery trip expect it to have the same taste and texture every time. Contamination of any sort, whether it be by foreign material or other microorganisms, is unacceptable. At best, this can break the customer's trust. At worst, it can lead to a product recall or food poisoning.

Consistently producing high quality natto therefore requires strict adherence to Good Manufacturing Practices (GMPs), food sanitation and safety programs such as Hazard Analysis and Critical Control Points (HACCP), as well as inherently good design of the facility and the equipment contained within it.

The contents of this chapter will describe the entire natto making process, the design of equipment and facilities, and some of the production measures and programs needed to ensure product safety.

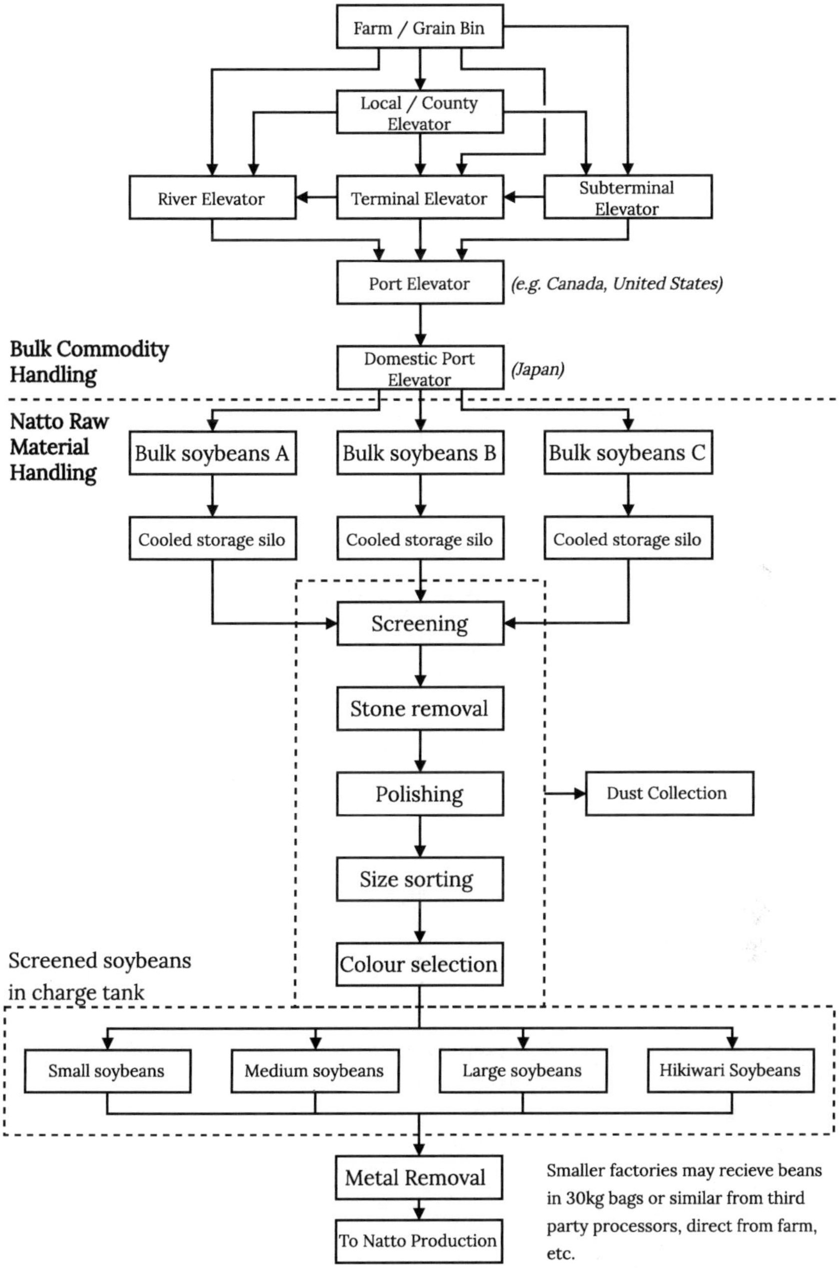

Figure II.1 Soybean raw material handling process (example).[22,23]

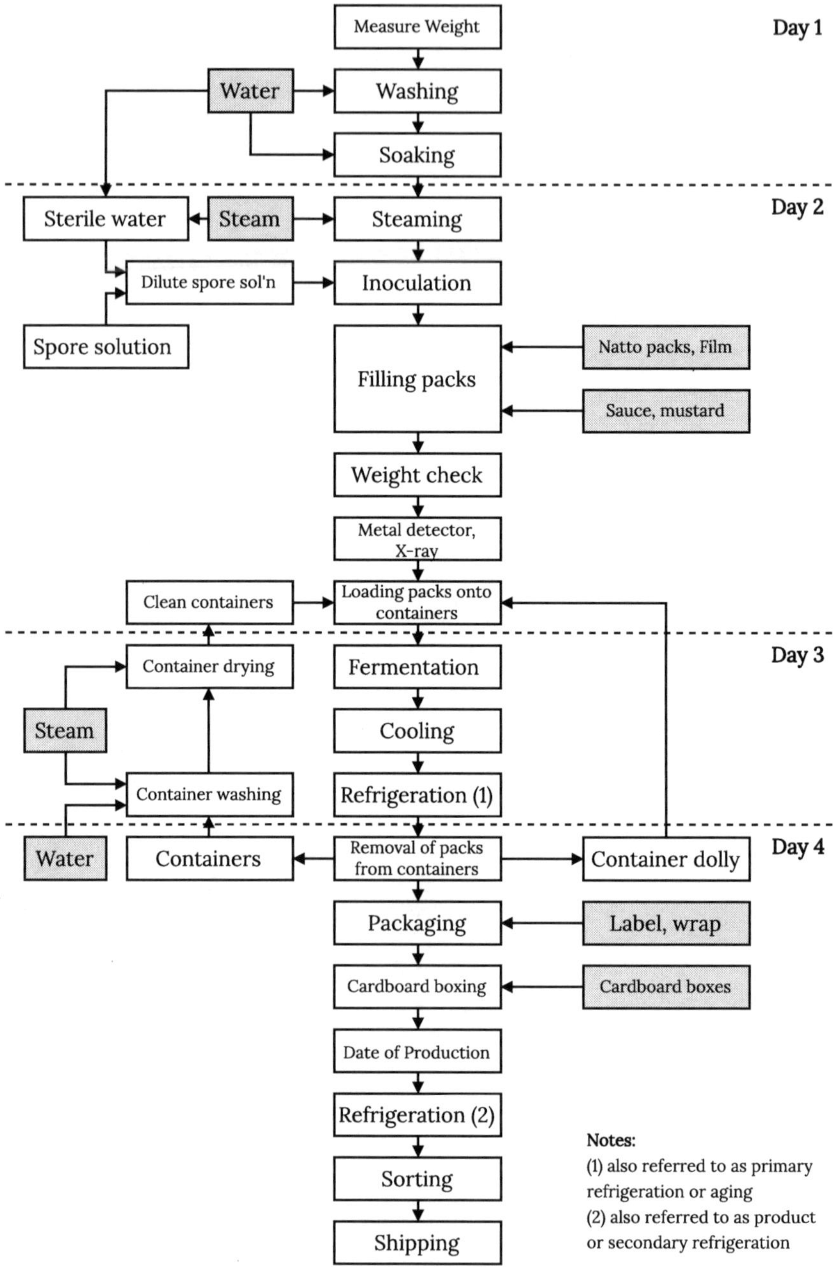

Figure II.2 Block flow diagram of a natto making process (example).[22]

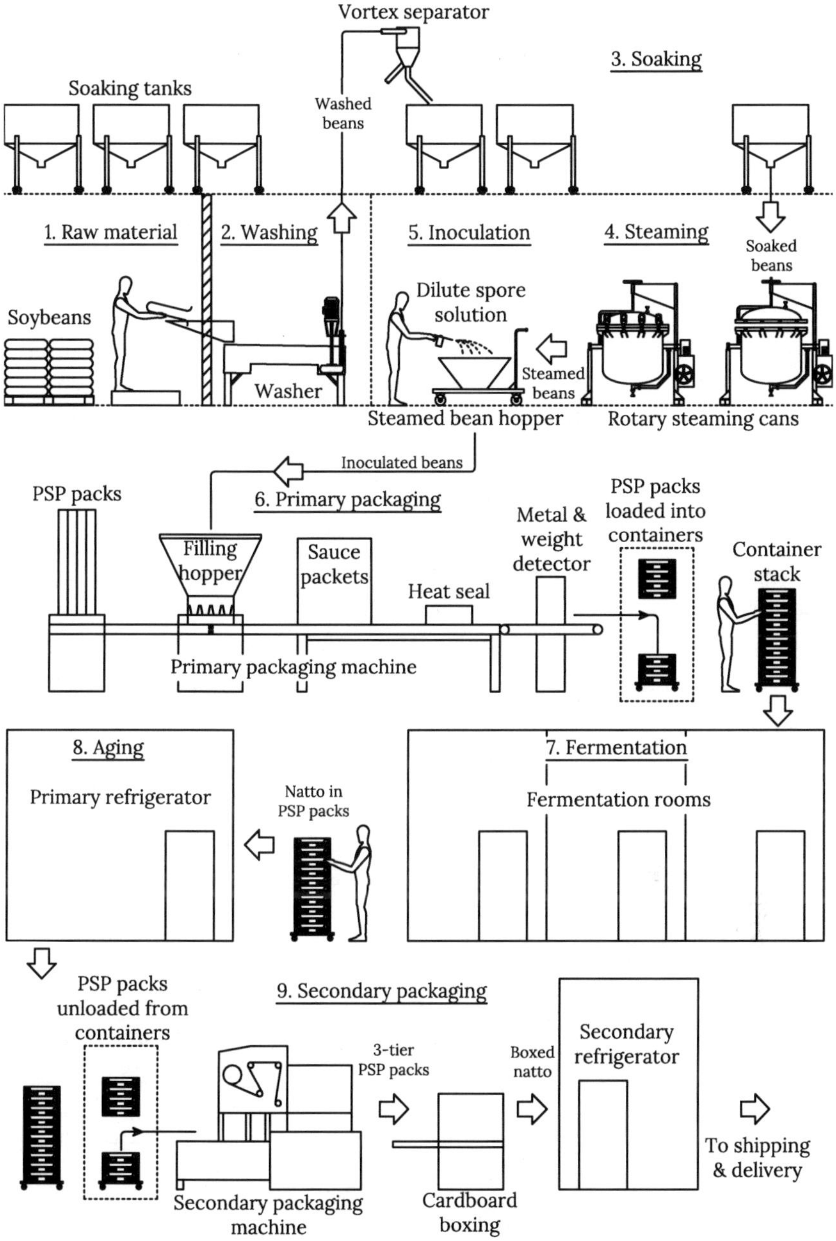

Figure II.3 Process flow diagram of a natto-making process (example).[22]

1. Natto Production Process

1.1 Raw Material Handling

Dry Soybean Handling

Soybeans harvested from the field must be stored at an appropriate temperature and moisture content to preserve their quality. High temperature and high moisture causes the beans to deteriorate, and the two factors are synergetic.[24]

Soybeans harvested in the autumn are typically stored in grain bins, which are equipped with a false floor and a blower that uses ambient air to dry the soybeans. A moisture content of under 14% is desirable, which can be achieved by drying at 17°C and 70% relative humidity, or lower.[24]

The bulk soybeans are then shipped to an intermediate grain processor before reaching the natto factory. Smaller operations might skip the intermediate processor altogether, purchasing directly from farms, while large natto factories will process bulk grain deliveries themselves.

The grain processor will remove as much foreign debris as possible, polish them (dust and soil removal), sort by size, and perform metal detection and removal of discoloured beans. For smaller processors, the beans are then packaged into 30kg paper bags and loaded onto pallets for sale to natto companies. For larger producers, beans are delivered in bulk and stored in silos at the factory. To preserve the beans' quality at this stage, producers store the beans at 15°C and a relative humidity of 60% or lower.[25]

Bean Size

Table II.1 Natto size grades based on the raw soybean size.[21,26,27]

Size Grade	Screening Mesh Size (Minimum Diameter)	National Natto Cooperative Voluntary Standards (Diameter)
Extra Large (極大粒)	≥8.5mm (Not official)	Not an official size class
Large (大粒)	≥7.9mm	>70% of beans >7.3mm
Medium (中粒)	≥7.3mm	>70% of beans 6.4–7.3mm
Small (小粒)	≥5.5mm	>70% of beans 5.8–6.4mm
Extra Small (極小粒)	≥4.9mm	>70% of beans <5.8mm

Soybeans for natto are graded by size, based on their ability to pass through progressively finer mesh sizes in a sorting screen. The size grade is displayed on a package of natto, and small grain soybeans are the most common.

Farmers will often use the 100–grain weight instead, which is the total weight of 100 individual beans sampled randomly. Under this metric, small grade natto soybeans weigh approximately 8–11g/100 beans. By comparison, generic food grade soybeans (those destined for miso, shoyu, tofu, soy milk, etc.) weigh approximately double, at 17–26g/100 beans.[28] These numbers vary by supplier, strain of soybean, and growing conditions.

Soybeans with a high surface-area-to-volume ratio and a high concentration of soluble sugars tend to promote the growth of *Bacillus subtilis*. However, soybeans properties can vary considerably by growing year, and each factory can tweak the production process as required to accommodate beans of different cultivars, sizes, and compositions.

Table II.2 Natto grade and generic food grade soybeans sampled in Canada, 2019–2020.[29]

Parameter (Average)	Natto Grade		Generic Food Grade	
Year of Harvest	2020	2019	2020	2019
Physical Parameters				
100-grain weight (g/100)	9.0	9.8	20.6	19.7
Water absorption ratio	2.21	2.24	2.13	2.14
Chemical Composition (g/100g dry matter)				
Protein	39.0	39.0	41.5	41.2
Oil	20.7	20.0	21.1	20.8
Sugars (g/kg dry matter)				
Sucrose	54.6	76.2	64.6	76.8
Raffinose	5.5	7.6	6.4	7.6
Stachyose	39.5	35.7	27.7	33.7
Verbascose	1.2	1.7	0.4	1.2

Washing

The soybeans are then washed to get rid of any remaining foreign substances, such as pebbles, dust, grease, and dirt, and thoroughly remove any surface bacteria. The standard is to wash the beans down to a bacterial count of 10^2 colony forming units per gram (CFU/g), although this isn't a parameter measured in normal production.[30] At small scales, washing is done by manually by swirling beans in a bucket with tap water, then straining.

At large scales, multiple types of bean washers are available. Screw-type washers transport beans upwards while heavier materials sink to the bottom. Drum-type washers tumble the beans inside of a mesh drum, set inside a larger drum partially filled with water. Rectangular washers pump the water and beans across a sloped basin, which separates contaminants by density. All of these washers operate on the basis that, in a moving mixture of water and beans, heavier materials such as pebbles will sink to the bottom, and light materials such as dust and oils will float to the top.[30]

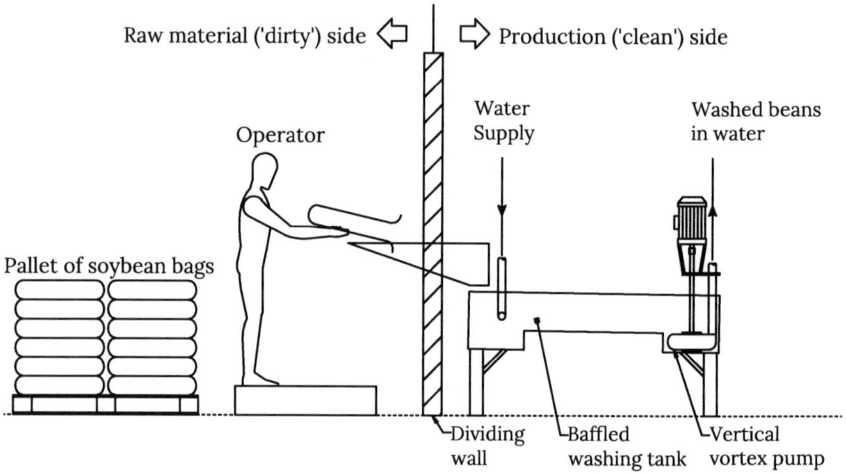

Figure II.4 Soybean washer (rectangular baffled tank)

At this stage, the washed, unsoaked beans are durable, and they can be transported by pumping the beans with water via a centrifugal pump. Pumping systems are equipped with vortex separation devices at the receiving end, which separates out the beans from the conveying water, allowing the water to be reused.

Soaking

The soybeans are soaked to allow the beans to fully absorb water to facilitate the cooking process. Soybeans are soaked until they are 2.2–2.4 times their dry weight, but wider ranges have been recorded depending on the bean's composition and initial moisture content.[25,30] A fully soaked bean, when split in half along the two cotyledons, is flat with a uniform colour and no voids.

Table II.3 Typical soybean soaking times

Soaking Temperature	Soaking Time	Notes
Whole Soybeans (Small Grain) - from Watanabe[31]		
5°C	24h+	Refrigerated soaking
10°C	23–24h	Generally accepted best practice
15°C	17–18h	
20°C	13–14h	Lactic acid bacteria may proliferate
25°C	7–8h	
Whole Soybeans (Small Grain) - from Hasegawa[31]		
10°C	24h	
20°C	16h	
Split Soybeans (Hikiwari) – from Watanabe[32]		
10–15°C	2.5h	Depends on particle size

Soaking proceeds faster the higher the temperature. In general, large soybeans take longer to soak, whereas split soybeans for *hikiwari natto* soak rapidly.

However, soaking at higher temperatures can allow lactic acid bacteria to proliferate. Although these bacteria will be killed during subsequent pressure cooking, acidic metabolites like lactic acid can reduce the pH of the beans, inhibit the growth of *Bacillus subtilis*, and in extreme cases, completely halt their growth. Excessive soaking time can also leach out the bean's soluble sugars, which are consumed by the lactic acid bacteria, reducing the amount of sugars available to *B. subtilis*.

The most common practice is to soak soybeans with tap water at, or pre-cooled to 10°C, and allow the soybeans to soak at room temperature. The water temperature will gradually rise during the soaking period.

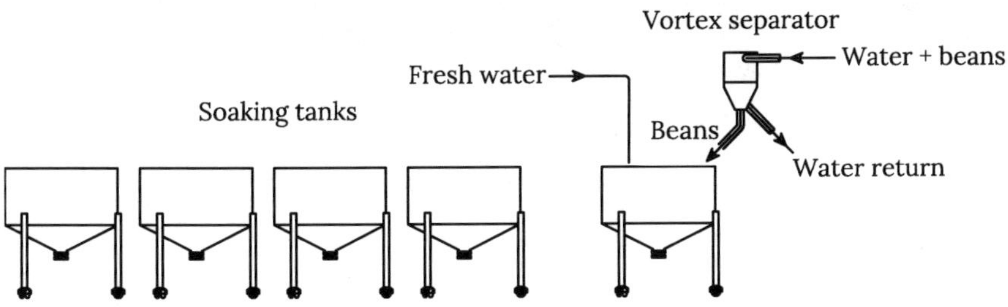

Figure II.5 Soaking tanks and vortex separator

Photo II.1 Unsoaked soybeans

Photo II.2 Soaked soybeans

1.2 Steaming, Inoculation and Filling

Pressure Steaming

Fully soaked beans are drained of water, sometimes for up to 30mins–1h,[31] then loaded into steaming vessels. Steaming is done with a rotary pressure steaming can connected to a separate, natural gas fired boiler. Without additional pressure, intact soybeans are notoriously difficult to cook, and take between 4–5 hours to steam at 100°C and atmospheric pressure.

Cooking under pressure saves a significant amount of energy, since it increases the boiling temperature of water and allows the soybeans to be cooked at between 121°C–132°C and pressures of 1.5–2.0kg/cm^2 (147–196kPag), reducing the cooking time to under one hour.[25] Pressure steaming was quickly adapted in the modern natto-making process as soon as the Japanese had access to the required equipment and manufacturing techniques.

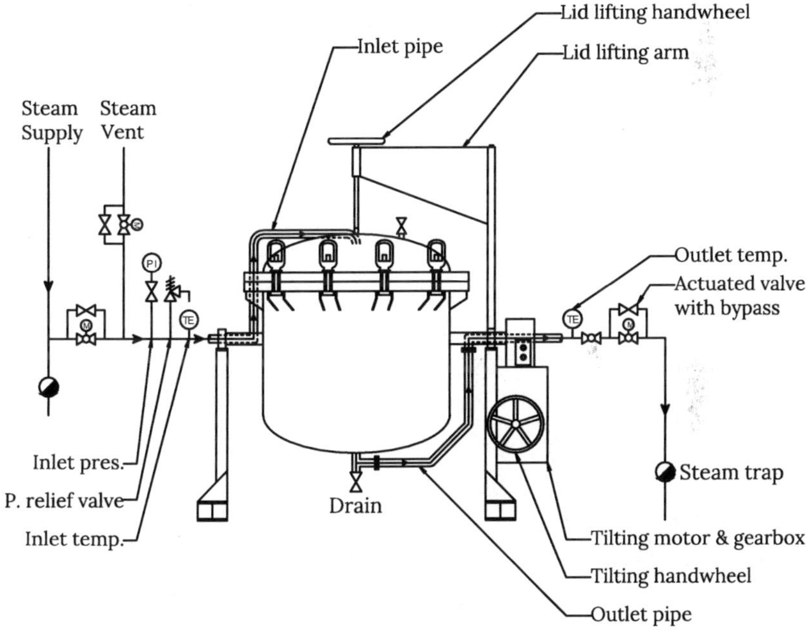

Figure II.6 Automatic rotary steaming can

Even with pressure cooking, these early methods were somewhat inefficient because they used direct-fired pressure steaming cans, where the heat source (gas, oil, wood, or coal) was directly lit below the cooker. This is comparatively inefficient because direct fired cookers allow heat to escape through the sides of the cooker and the stove below.

In any natto facility of a significant scale, a dedicated boiler system is required. These natural gas boilers are highly efficient, using multiple passes of tubes to generate steam, which is then piped to the pressure cooker. The steps to steaming are as follows:[25]

1. Beans are loaded into the cold steaming can, which is sealed shut through a ring of hinged, captive bolts around the flange of the cooker. A drain valve located at the bottom of the cooker is kept open.
2. Steam is sent into the top of the pressure cooker from the boiler. The steam condenses on the cold beans, and condensate is discharged out the bottom of the cooker. This lasts for around 10–20 minutes.
3. Once the pressure cooker and beans reach temperature, only steam will exit the drain valve. At this point, the drain valve is closed and the pressure inside the cooker begins to rise, reaching 1.5–2.0 kg/cm^2 within 10–20 minutes.
4. The beans are cooked at the target pressure for 30–40 minutes. Excess condensate is removed continuously through an outlet valve.
5. Once the cooking time has elapsed, the supply of steam is shut off. The pressure cooker is allowed to depressurize naturally, while all valves are still closed, for 10–15 minutes.
6. Finally, the bottom valve and vent valves on the steam pipe are opened, allowing the pressure to equalize inside and outside the cooker, and then the lid is unbolted and opened.

Adjustments are made to the steaming process depending on a variety of factors. Low pressure steaming ($1kg/cm^2$) can take around 60 minutes and high pressure steaming ($2kg/cm^2$) takes around 30 minutes once the pressure cooker reaches temperature.[31] Beans soaked in colder water, especially in the winter, take longer to reach temperature during the initial steps.

The cooking time, temperature and pressure are determined by the quality of the resulting steamed beans. The steamed soybeans should be soft and slightly sweet, indicating the presence of simple sugars. They should also be darker brown in colour compared to soaked beans, which is caused by

Maillard reactions occurring at high cooking temperatures. The hot beans should be soft enough to be crushed by squeezing between the thumb and ring or pinky finger.[31] There is also a quantitative test for the bean's hardness: allow the beans to cool to 30°C and press them on a weigh scale. The average crushing force of a single steamed, cooled bean should be around 120g.[31]

The cooking process can be automated or manual. In manual operations, an operator must open and close valves according to the time, temperature, and pressure of the steam system and pressure cooker. In an automated plant, these actions are performed by a computer or programmable logic controller (PLC). Large facilities run multiple pressure steamers, allowing a boiler to be kept continuously running, and steam can be sent into different cookers in succession. In smaller facilities, the boiler is only fired on for the cooking process.

The steaming process produces a soybean broth, which is currently discarded as wastewater and can be a significant expense for a natto facility. Nevertheless, it is currently too expensive to treat this wastewater prior to disposal. Natto researchers are always on the lookout for a cost-effective wastewater treatment process, or a cooking process that doesn't produce any wastewater.[22,30]

Batch type pressure cooking is also inefficient compared to continuous pressure cookers, like those used on defatted soybean meal in large *shoyu* factories. However, the production of natto requires intact soybeans, which is not possible with *shoyu*-type continuous pressure cookers.[22]

Inoculation

Inoculation of the steamed soybeans is done with a prepared solution of *Bacillus subtilis* endospores (known as *natto-kin*). These spores, formed by the bacterium as a survival mechanism, are highly resilient and can survive temperatures up to 100°C and have a long shelf life.

There are three major spore producers in Japan:

1. **Miyagino Natto Manufacturing** in Sendai City
2. **Naruse Fermentation and Chemical Research Institute** in Tokyo
3. **Yuuzo Takahashi Laboratory** in Kamiyama City

Spores are sold in concentrated liquid suspensions in sterile water, or solid suspensions on powdered silicon dioxide (sand).

The standard inoculum amount is 5×10^3 colony forming units (CFU) per 1g of cooked soybeans, but in bench scale tests it can be as high as 5×10^4 CFU/g.[31] The general rule of thumb is that 5mL of concentrated spore solution from the supplier is used for every 60kg of dry soybean equivalent (roughly 132kg steamed soybeans). The strength of the spore solution, and recommended dosage per kg of dry soybeans, is provided on the package label.

This concentrated spore solution must be diluted to effectively disperse the spores onto the surface of the steamed soybeans. To do this, hot water is taken from the boiler and allowed to cool to 50°C for safe handling. The concentrated spore solution is then added to the cooled, sterilized water and mixed.

If using a pressurized sprayer, around 1L of dilute spore solution is used to inoculate 60kg of dry soybean equivalent. If using a watering can, around 1.25L of spore solution is used for the same amount of soybeans.[25] The inoculation is typically performed right at the hopper used to unload the pressure cooker, in the pressure cooker itself, or during the unloading process. Simply unloading the pressure cooker mixes the beans and spore solution, which eliminates the need for any dedicated mixing devices like mixing drums, which were originally used in Jun Hanzawa's method in the 1920s and 30s.

Inoculation with spores is performed at 70–90°C, with a target temperature of 85°C.[25,31] Inoculating at a high temperature minimizes the chance of contamination, since this temperature is too high for viruses or vegetative (non-sporulated) cells from potentially pathogenic bacteria to survive. Inoculating at a high temperature may also thermally shock the spores, allowing them to germinate quickly and grow once incubation temperatures are reached. However, this process is not well understood, and the effectiveness of 'heat shocking' endospores varies greatly by strain of *Bacillus subtilis*.[33]

Filling (Primary Packaging)

Steamed, inoculated soybeans are loaded right into its packaging while still hot, immediately after inoculation. The beans are loaded into a hopper, and a filling mechanism moves the soybeans down stainless steel channels into a dosing mechanism. The dosing mechanism dispenses the correct amount of steamed soybean by volume into each container using a pair of pneumatically driven slide gates.

For polystyrene paper (PSP) packages of natto, 40–50g of soybeans are dosed into each package. For paper cups, between 30–40g of soybeans are dosed into each cup. High-speed machines can fill between 120–240 individual pieces per minute.[25,30]

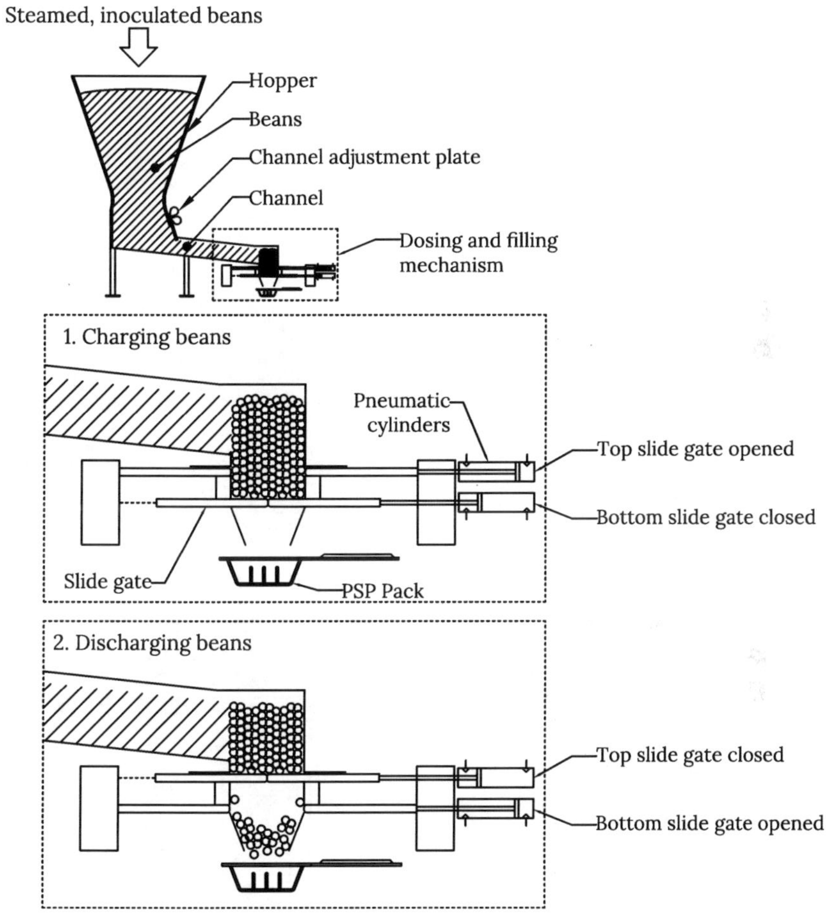

Figure II.7 Filling mechanism

After filling, a film is applied over the beans, on top of which packets of sauce are optionally added. Finally, the top lid of the PSP package is folded over the lower box and heat-sealed. For cup type containers, no further action is required after applying the film and sauce packets.

Packaging materials other than PSP packs and cups are usually filled by hand, since the bulky packaging material is difficult to handle, and too few factories use this type of packaging to warrant designing any specialized machines. Manual filling is labour intensive and contributes to the high cost of natto made using speciality packaging material. Manual filling is sometimes aided by a dosing machine that doses the correct volume of soybeans, but the package is still handled by the worker.

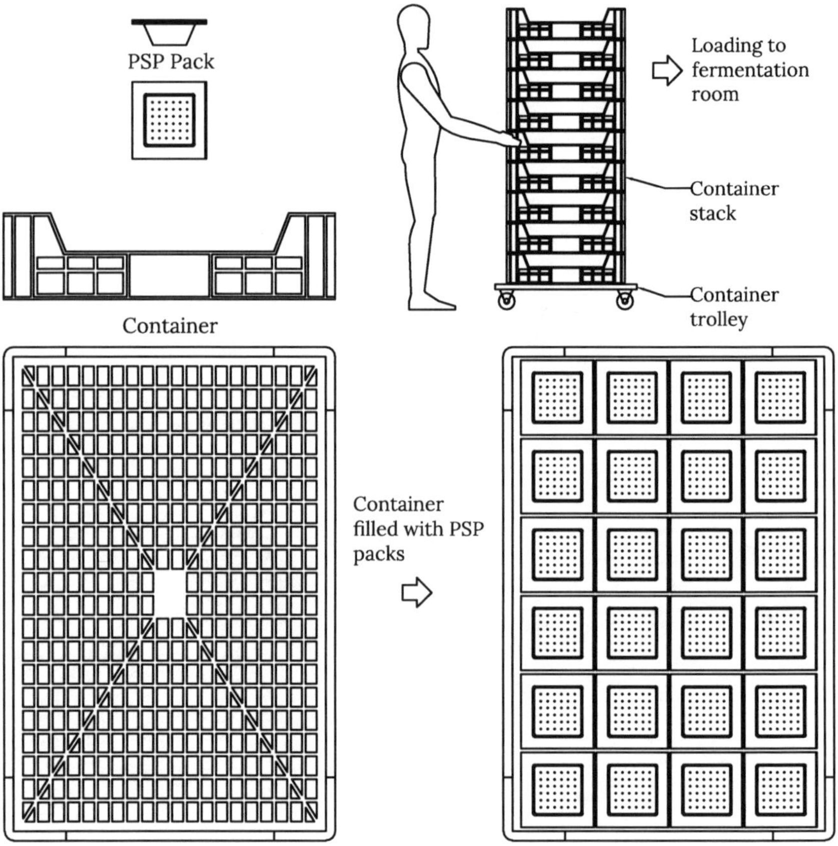

Figure II.8 Natto packs, containers, and container stacks

The filled containers are finally weighed to check for filling accuracy and passed through a metal detector or X-ray before being loaded into plastic trays. In natto literature, these trays are referred to as 'containers', and resemble the plastic, stackable dish and flatwater racks used in the commercial restaurant industry. Around 35–42 packages are loaded into each container, and 12–15 containers are stacked, loaded onto trolleys, and wheeled into the fermentation room.

Packaging Material Properties

Table II.4 Natto packaging properties.[25]

Properties during:	Package Filling			Fermentation			Refrigeration			Others	
Container Type	Machine suitability	Firmness	Ease of sanitation	Heat Retention	Water Retention	Airflow	Heat retention	Water retention	Firmness	Ease of consumption	Incineration
PSP packs	◎	◎	◎	◎	◎	△	◎	◎	○	○	×
Paper cups	◎	◎	◎	×	○	○	×	○	○	◎	◎
HDPE cups	◎	◎	◎	◎	◎	×	◎	◎	○	○	×
PVC cups	◎	◎	◎	○	◎	×	△	◎	◎	◎	×
PE lined *kyogi*	○	△	○	×	○	×	×	○	△	×	×
Parrafin lined *kyogi*	○	△	◎	×	○	×	×	○	△	×	×
Plain *kyogi*	×	×	○	×	○	○	×	○	△	×	◎
Sudare	×	×	×	○	○	◎	○	○	△	×	◎
Rice straw *(Tsuto)*	×	×	×	○	○	◎	○	○	△	×	◎

(◎ *Excellent,* ○ *Good,* △ *Mediocre,* × *Bad*)

The type and size of packaging has an important role in many aspects of natto making, from filling to fermentation to marketing. PSP packages are by far the most common, but other packaging options include paper and plastic cups, thin sheets of pine (known as *kyogi*), bamboo, rice straw, or reed sheets (known as *sudare*) and even the original rice straw bundles (known as *tsuto;* rice straw is known as *wara*), although rice straw is often steam sterilized rather than relying on spores present in the straw.

The choice of packaging material affects its compatibility with filling machines, and its permeability to heat, moisture, and airflow during fermentation and refrigeration. Plastic materials such as PSP and plastic cups tend to retain heat and water well, while plant-based materials are highly permeable to air and water. These factors affect the required humidity and temperature setpoints during fermentation and refrigeration.

Plant-based packaging materials also appear more premium than their plastic counterparts and fetch a higher price at the storefront. Commercial and institutional kitchens serving natto, such as those in restaurants and hospitals, also purchase larger packaging formats, such as 500g packs. *Hikiwari natto* is also available in squeeze tubes, which are loaded after fermentation.

1.3 Fermentation

Overview of the Fermentation Process

The objective of natto fermentation is to ferment the soybeans to a point where they yield a sticky biofilm that forms long, stretchy threads when stirred, and have the odour expected of natto. Properly made natto, after refrigerated curing, should not be overly bitter or have a strong ammonia smell: indicators that the fermentation has gone too long or at too high a temperature. Properly made natto should not have weak threads that look like cobwebs, and the natto should not have the aroma of regular steamed soybeans: indicators that the fermentation has gone for too short.

Bacillus subtilis is an aerobic bacterium and natto production requires oxygen for growth (note that technically *Bacillus subtilis* is a facultative aerobe, see Chapter IV). It lacks the ability to form filamentous hyphae, like those of fungi, and cannot burrow deep into the substrate to find more food. Instead, the bacteria reside on the surface of the soybeans, which must have sufficient moisture to prevent the growing colony from drying out.

B. subtilis grows best at 38°C. As the bacteria grows inside a package of natto, its metabolism will generate a significant amount of heat, causing the temperature to rise. If this is not managed, the fermenting soybeans will easily exceed 55°C, causing bacterial autolysis and halted growth.[25] This temperature rise will happen more rapidly if the natto container is well insulated, such as in PSP packs.

On the other hand, high quality natto is made when the soybean's temperature is allowed to reach around 52°C during the latter phases of fermentation, which produces better flavour. This is coupled with a reduction in humidity during the final stages of fermentation to concentrate the flavour of the natto and slightly reduce its moisture content.

The fermentation process is a balancing act of managing the temperature and humidity of the room, the temperature of the natto itself, and taking into consideration different packaging materials' differences in thermal conductivity, and permeability to air and moisture.

When describing the kinetics of fermentation in a batch environment such as a package of natto, the process is broken down into four 'phases' on a bacterial growth curve, describing the state of the bacterial population at that time.

In the **Induction** or **Lag** phase, the bacteria are germinating from their endospores, and starting to replicate.

In the **Logarithmic** phase (also known as the exponential phase), an abundance of nutrients allows the bacteria to grow at an exponential rate. For *Bacillus subtilis*, the population of bacteria can double every thirty minutes (also known as its doubling time).

In the **Stationary** phase, the population of live bacteria is maximized, and further growth becomes limited by the availability of food. The rate at which bacteria duplicate and the rate at which they die or sporulate reach an equilibrium, and the population stays around the same.

In the **Death** phase, the bacteria begin to run out of food and nutrients, and what remains can no longer sustain bacterial growth at the same rate as before. The bacteria begin to die and sporulate faster than they reproduce, and the vegetative population declines while many bacteria will form endospores.

In natto production, the main objective in fermentation is to halt fermentation towards the end of the stationary phase by forced **cooling**.

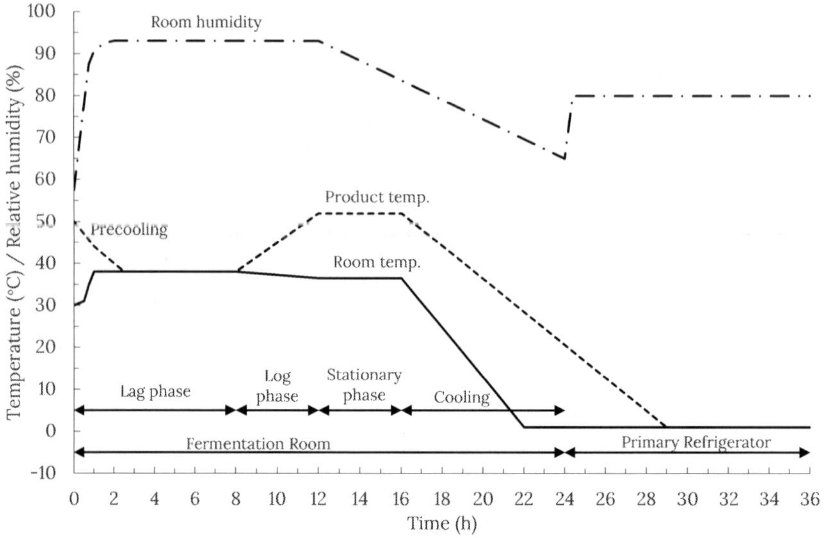

Figure II.9 Natto fermentation conditions

Loading the Fermentation Room

Fermentation rooms are equipped with multiple humidity sensors and temperature sensors for both the room and natto product, herein referred to as the **room temperature** and **product temperature**, respectively. The room's mechanical functions include heating, cooling (by ventilation or refrigeration), humidification, dehumidification, and the control of supply and exhaust air. A control panel mounted near the front of the room determines the setpoints for each stage of the process, and records data for quality control. This panel adjusts individual parameters for each stage of the growth cycle.

Induction (Lag) Phase (0–8h)

Trolleys containing stacked containers of inoculated soybeans packs are pulled into the fermentation room while still warm, at around 50°C. Between 0–1h, the pre-cooling phase occurs, and the product and room temperature reach an equilibrium between 37–40°C.

The first 8 hours of operation (0–8h) are known as the induction phase. Humidity is kept high at between 90–93% RH during this period to prevent the surface of the soybeans from drying out. Within the first 2h, the spores of

B. subtilis germinate, and then begin to propagate on the surface of the soybeans.[25]

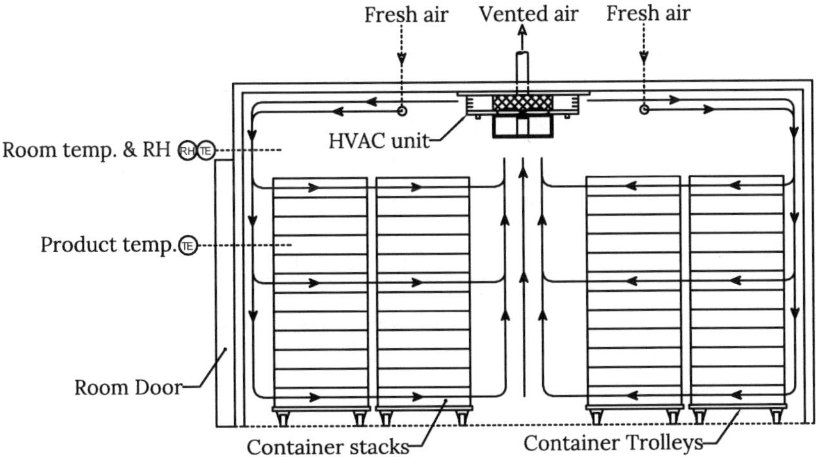

Figure II.10 Fermentation (incubation) room

Logarithmic Phase (8–12h)

The next 4 hours (8–12h) are known as the logarithmic growth phase. Here, the growth of *B. subtilis* is rigorous and exponential, with bacterial counts doubling around every 30 minutes. Enough metabolic heat is generated by the bacteria during this period to cause the product temperature to rise relative to the room temperature. The rate of temperature rise is affected by the soybean's surface area (high surface area results in rapid growth and high rate of temperature rise) and the properties of the container (a highly insulated container results in high rate of temperature rise). By the end of this period, the product temperature will reach 50–52°C.[25]

Stationary Phase (12–16h)

The next 4 hours (12–16h) are known as the stationary phase. The general best practice is that a product temperature of 50–52°C must be maintained for around 4h for good quality natto.[34] The exact biochemical mechanisms behind this target temperature are not fully researched, and are instead based on qualitative assessments of the natto produced at different temperature profiles. If the product temperature exceeds 55°C, the bacteria will undergo autolysis, which compromises the natto's quality and produces off-flavours and textures.

During this phase, the room temperature and room humidity setpoints are dependent on the properties of the natto container. For well-insulated containers, such as PSP, there is a large difference in the room temperature and product temperature, so the room temperature is lowered slightly to around 36°C, to achieve a stable target product temperature of 50–52°C.[25] On the other hand, a permeable container such plastic or paper cups may require an increase of room temperature to achieve the same product temperature, depending on the size of the soybean and the density of containers inside the room.

One of the reasons square polystyrene containers are so popular with natto manufacturers is that they require comparatively lower heating requirements for the natto room. Grooves molded into the inner surface of the container also allow for greater airflow around the beans.

Room humidity is reduced during the stationary phase, from around 90% RH to 80% RH by 16h. The reduced RH allows heat to dissipate from the natto containers through evaporative cooling, and starts to reduce the natto's surface moisture, concentrating its flavour. This reduction in moisture will continue towards the last phase of fermentation and during refrigeration.

Cooling Phase (16–24h)

The last 8 hours of operation (16–24h) are known as the cooling phase. The room temperature is reduced, with the target room temperature of 5°C around 22h. The product temperature lags the room temperature, and reaches a temperature of 20–30°C at 22h.

The humidity is continually reduced during this period. Dehumidification continues to concentrate the substances on the surface of the soybean.

By the end of the cooling phase, some bacterial autolysis has occurred. These bacteria break down and spill their intracellular components, including enzymes and metabolites, into the natto, contributing to its flavour. Enzymes will continue to act slowly on the natto during refrigeration. Some bacterial cells also form endospores, and the distribution is roughly 10–40% endospores and 60–90% vegetative cells by count at the end of fermentation process.[35]

Photo II.3 Finished Natto.
Note the wrinkled biofilm; smooth sections are where the top film touches the beans.

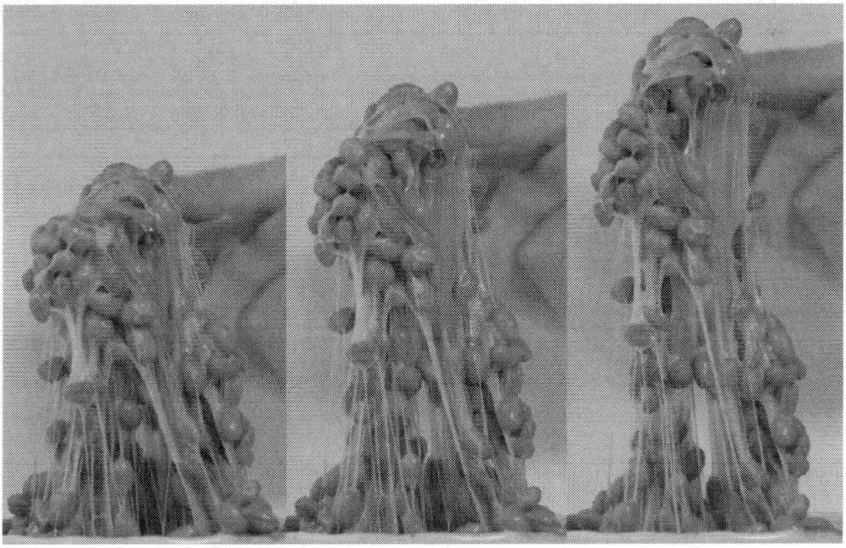

Photo II.4 Finished Natto, stirred and lifted.
Threads should be sticky and mucilaginous as shown above.

1.4 Post-Fermentation and Distribution

It's important to halt the growth of natto bacteria at the correct time to prevent the over-production of ammonia. Natto bacteria prefer soluble sugars as their carbon source, but once these are exhausted, the bacteria will increasing rely on amino acids as a carbon source. The resulting deamination reaction produces ammonia, which has a pungent smell.

During the next 12h, and continuing throughout the distribution cold chain, the natto is allowed to mature. Although the bacteria are mostly inactive, their enzymes continue to work slowly within the natto, further developing its flavour. Off-gasses and small amounts of ammonia produced during the fermentation process are also volatilized, improving the natto's odour. The packages of natto lose a small amount of moisture during refrigeration, and in some cases, the biofilm on the surface of the beans will go from a wrinkled, thick layer to a thin film resembling frost.

Primary Refrigeration, Secondary Packaging and Refrigeration

The stacked containers of packaged natto are pulled out of the fermentation room and moved into a refrigerated room for primary refrigeration. Here, they are allowed to fully cool to 0–5°C. Once cooled, the stacked containers are removed from the primary refrigeration room and wheeled to a machine that removes the packages of natto from their plastic containers. The containers are then cleaned and returned to the primary packaging process.[20,25]

Individual packages of natto are now sent into secondary packaging. Here, the final label is applied. Most PSP packs are stacked in 2–tier, 3–tier, or 4–tier configurations and wrapped with a single label. Paper cups are usually packaged in 3's and shrink-wrapped along with their label. Then, the natto packages are loaded into cardboard boxes and moved into a refrigerated room (secondary refrigeration) to await shipping. Producers usually require that the natto spends at least 24h under refrigerated conditions to mature before reaching the customer.

Distribution

Quality control is conducted after fermentation, and after primary and secondary refrigeration. This evaluation consists of both qualitative and quantitative tests, the details of which are described in Chapter III.

The natto is kept below 10°C throughout the supply chain. Refrigerated trucks are required for distribution to grocery stores, where they are found for around 100 yen for 3x50g packages. PSP packs are by far the most common form of packaging in large chain supermarkets.[25]

The best before date is voluntarily determined by the producer, but is generally 9 days after the date of quality inspection.[30]

Old Natto

The nature of 'best before' dates is that they are simply an estimate; there is no exact amount of time after which a package of natto goes from edible to inedible. However, with extended storage at refrigeration temperatures, the quality of natto degrades in one of many ways.

Firstly, the proteolytic enzymes of natto continue to break down soybean proteins at refrigeration temperatures. This process produces more amino acids within the natto, leading to a change in flavour. One of these amino acids, tyrosine, is abundant in soy protein and has a low solubility, so it tends to precipitate into small crystals. Although harmless, these tyrosine crystals can create a grainy or mealy texture.

Long-term storage of natto can also result in the formation of struvite, which is a mineral consisting of ammonium, magnesium, and phosphate. The concentration of ammonium increases as the bacteria continues to deaminate amino acids, while magnesium and phosphates are released from their bound forms as components of the bacteria and soybean degrade over time. Struvite is harmless when ingested in such low quantities, but also contributes to a grainy or mealy texture.

Both tyrosine and struvite formation are exacerbated by the continued loss of moisture in the refrigerator. Extended storage inside a refrigerator also dries out the natto, leading to a harder texture.

Storage under warmer temperatures also revives bacterial activity, and risks contamination with other microorganisms, especially if the container is opened and the protective film is removed.

Photo II.5 White tyrosine crystals formed on the surface of old natto

1.5 Process Optimization for Different Beans

The method described prior applies to small grain natto, but adjustments to the process can be made depending on the cultivar and size of bean. This requires some trial-and-error for different beans, but in general, larger beans require longer soaking, steaming, and incubation times, and have a slower rise in temperature during fermentation. The opposite is true for smaller beans.

Thinner seedcoats, or beans processed to remove the seedcoat (like *hikiwari natto*) also results in faster growth, and a shorter incubation time. Beans with a higher fraction of soluble sugars also support more bacterial growth and stronger stickiness. Details of the mechanisms and biochemistry of this process are explained in Chapter V.

1.6 Hikiwari Natto

Hikiwari natto is made by breaking large grain soybeans into smaller pieces and removing the seed coat. Doing so provides more surface area to facilitate the growth of natto bacteria, and large grain soybeans are cheaper than small grain soybeans.

Hikiwari natto is thought to be developed in the Akita region, where in the traditional farmhouse setting, soybeans were first roasted in a pan (like for *kinako*, roasted soybean flour, or *iri-mame*, roasted soybeans). The roasted, cooled soybeans would then be broken up with a mortar and pestle.

Because there is no seed coat, *hikiwari natto* is comparatively easy to eat, making it suitable for infants and the elderly. It's also more amenable to mechanization since the beans don't need to be kept intact throughout the production process. This makes it popular for *maki* and *gunkan-maki* produced by convenience stores, as well as for *natto-jiru*, school lunches, and hospital meals.[32]

Cracking Beans

Modern methods for making *hikiwari natto* involve splitting the soybean in one of two ways: with and without heating. In the former, the dry soybeans are heated to around 85°C to reduce the moisture to under 11%. This process takes between 2–3 hours, after which the beans are cooled down. Removing moisture makes the beans brittle, and they are subsequently cracked into 2, 4, or 8 pieces per bean by a pair of rollers, depending on the product specification. This is done by adjusting the space between the rollers. The seed coat is removed with an air separator, and dust is removed with a sieve.[32]

Smaller factories split the soybeans without heating. This forms cracks that are rougher than with heated beans, but additional heating equipment is not required. Some factories will also run the cracked beans through a set of vibrating screens to further sort the fragments by size and remove any debris and small particles.[32]

Soaking Cracked Soybeans

With the cotyledon of the soybean exposed, subsequent steps proceed differently from whole grain natto. Firstly, cracked soybeans allow for a lot of the bean's internal components to be leached out into the soaking water, and soaking proceeds much faster.

Some producers adapt by soaking in a 0.3% salt solution in order increase the osmotic pressure of the soaking liquid. Soaking takes only 2.5–3 hours, and the beans are thoroughly drained after soaking for around 30 minutes.[32]

Another method used by producers is to limit the amount of soaking water, such that the cracked soybeans absorb all the soaking water with no excess. In this method, soybeans are loaded into a rotating, airtight container, into which hot water at 70–80°C is added at approximately 80% of the crushed soybeans by weight. All of this water is absorbed.[32]

Steaming

Steaming for *hikiwari natto* is different than whole grain natto. Firstly, in large facilities, a continuous steamer can be used, since there's no need to keep individual beans intact.

If a rotary can type steamer is used, a set of stacked strainers are needed to prevent the pieces of cracked soybeans from compressing into a lump under their own weight, which can lead to channeling steam around these lumps, and uneven cooking. Steaming proceeds at lower pressures, typically in three stages: 1kg/cm^2 (120°C) for 15 minutes, 0.6kg/cm^2 (113°C) for 20–30min, then atmospheric steaming for another 10–15 minutes.[32]

The inoculation procedure is the same as for whole grain natto.

Fermentation

Fermentation proceeds faster with *hikiwari* natto than for whole grain natto, since there is a greater surface area, and the insides of the soybean are fully exposed. The rise in temperature during the logarithmic period will be steeper, and once the stationary period is reached, the product temperature is kept at 49–50°C for around 3h. After 15h from the start of fermentation, the room temperature will be lowered, and then *hikiwari natto* is removed from the fermentation room between 18–20 hours.[32]

The remainder of the natto-making process, including primary refrigeration, secondary packaging, and secondary refrigeration, are the same for both *hikiwari* natto and whole soybean natto.

1.7 Batches and Record Keeping

Batch Size

Throughout the production process, an accurate record of each production step is crucial for product safety, repeatability, and troubleshooting. The entire process is defined by individual batches, and each unit process is sized to handle one batch of dry soybean equivalents, usually in units of 'bales'.

A 'bale' (*hyo*, see Glossary of Japanese Terms) is a historical Japanese unit of volume corresponding to 60kg of dry soybean equivalent, similar to the western unit 'bushel'. For example, a factory might be designed to work in batches of 2 bales, or 120kg:

1. Each soaking tank can soak 120kg of dry soybeans.
2. Each steamer can steam 264kg of soaked soybeans (120kg x 2.2 water absorption ratio).
3. This steamed batch generates enough beans for 4800 individual PSP packages of natto (264,000g / 55g per pack).
4. The 4800 packs are loaded 40 packs to a container, 10 containers to a stack, and 12 stacks are moved into a single fermentation room.

This entire production batch would be assigned a **lot number** for record keeping, and every parameter of the production process would be recorded. This record is a crucial component of GMPs, and in the rare case of a recall or customer complaint, the lot number helps back track the process and identify other items in the batch that need to be recalled. An example batch record, used by natto-producers in Japan and translated to English, is provided below.[25]

Table II.5 Sample Batch Record

Tracked Parameters *(Equipment used)*	***Key Parameters & Discussion***
1. Raw Material (Soybean washer model no.)	
1. Variety 2. Place of origin 3. Production year 4. Grain Size: Large Medium Small Very small	

***Tracked Parameters** (Equipment used)*	***Key Parameters & Discussion***
Hikiwari 5. Purchase date quantity 6. Type of shipment Bulk Woven bags Paper bag 7. Supplier 8. Price 9. Contamination 10. Damage 11. Composition	
2. Soaking	
1. Tank number 2. Soaking quantity 3. Start of Soaking Water temperature Room temperature 4. Soaking time Start End 5. After Soaking Water temperature Room temperature 6. Water quality	Results and preventative measures Water absorption rate Germination rate
3. Steaming (Pressure steamer model no.)	
1. Steaming quantity 2. Steam source pressure 3. Steaming process Start Steam blow-out Reached pressure Pressure hold Depressurization Total time 4. After Steaming Soybean hardness (hot) Soybean hardness (cooled)	Results and preventative measures Hardness test
4. Inoculation	
1. Type of Bacillus subtilis natto 2. Quantity used 3. Amount of diluted solution 4. Inoculation method	Results and preventative measures Check starter (viability)

Tracked Parameters (Equipment used)	*Key Parameters & Discussion*
5. Filling (Filling equipment model no. and capacity)	
1. No. of packages / items	Results and preventative measures
2. Package type	Check for bacteria (plate count)
3. Package weight	
4. Filling time	
Start	
End time	
5. Room temperature	
6. Loading into Fermentation Chamber (Fermentation equipment model no.)	
1. Container (tray) type	Results and preventative measures
2. Fermentation chamber number	
3. No. of packages / items	
4. Package quantity per container (tray)	
5. Number of container (tray) stacks per cart	
6. Total number of containers / room	
7. Total loading	
No. of packages	
Soybean weight	
8. Loading time	
Start	
End time	
9. Room temperature during loading	
7A. Fermentation -Planning	
Fermentation chamber setpoints:	
1. Start of fermentation	
2. Pre-cooling (Temp, time)	
3. Lag (Temp range, time)	
3. Growth (Temp range, time)	
3. Stationary (Temp, time)	
6. Cooling (Temp, time)	
7. Oxygenation and aeration	
Duration & timing, times/hour	
8. Humidification	
Humidity setpoints/time	
9. Removal from room (time)	
10. Room temperature	
(outside of fermentation room)	
7B. Fermentation-Results)	
1. Start time of fermentation	Tasting
2. End time of fermentation	Examination of the results
3. Total Fermentation Time	Measures to be taken
4. Room temperature record	pH

Tracked Parameters (*Equipment used*)		*Key Parameters & Discussion*
	5. Product temperature record 6. Humidity record	Moisture content Ammonia-N
8. Secondary Packaging (Secondary packaging model no. and capacity)		
	1. Secondary packaging format 2. Start time 3. End time 4. Total Time 5. Packaging room temperature	Results and preventative measures
9. Refrigeration and Shipping (Refrigerator/freezer model no.)		
	1. Temperature setpoint 2. Time of loading 3. Time of removal 4. Total refrigeration Time 5. Shipping Product temperature 6. Cold Chain Vehicle type and temperature (List all vehicles)	Tasting (during shipping) Sales department evaluation Customer feedback pH Moisture content Ammonia-N
10. Cleaning and Sanitation		
	1. Objects to be cleaned and sanitized 1) Rooms Soaking room Steaming room Filling room Fermentation room Packing room Refrigerator Others 2) Machinery and equipment Soaking tank Steamers Inoculating equipment Filling machine Lifting hopper Filling equipment Filling table Container Cart Others 2. Cleaning and Sanitation Methods 1. Cleaning Type of cleaning agent	Results and preventative measures Check for bacteria (plate count)

Tracked Parameters (Equipment used)		*Key Parameters & Discussion*
	Concentration Amount of liquid used Temperature Rinse and wash 2. Sanitation Hot water sterilization Type of disinfectant Concentration Temperature Amount of liquid used Rinse & Wash	

Batch Timing

Natto is a perishable food that is produced and distributed on a continuous, daily basis. Each individual batch is produced across a total of 3–4 days, with the final product reaching store shelves on Day 4 and Day 5, according to the following schedule.[25]

Table II.6 Example Batch Schedule

Day	Production Steps
1	Raw material handling → Washing → Soaking starts
2	Draining → Steaming → Inoculation → Primary Packaging → Fermentation Start
3	Fermentation End → Secondary Packaging → Refrigeration (Primary)
4	Refrigeration → **Product ready for sale** → Cold Chain Distribution → Storefront
5	Cold Chain Distribution → Storefront (second day of inventory)

It's the job of every natto producer to deliver consistent batches of high quality natto, without any hazards or inconsistencies. This depends on a few factors such as facility design, food safety, and quality control, which will be discussed in later sections.

2. Facility Design

2.1 Equipment Suppliers and Natto Manufacturers

For most Japanese producers, the design of a natto facility doesn't fall on the factory owner or operator, but on the equipment manufacturer. Suzuyo Kogyo Co., Ltd., is the dominant player in natto equipment supply, and accounts for approximately 90% of natto equipment sales in Japan.[20] This company can deliver a turnkey factory, including sales planning, design, equipment manufacturing, testing, training, and maintenance. The other notable equipment supplier is Miwa Seiki Co., Ltd., which manufacturers equipment for filling packs and handling containers.

Overseas Manufacturers

Outside of Japan, producers don't have access to the same natto production infrastructure. Importing equipment is possible, but it's unclear whether large overseas natto producers like those in Korea have opted to use imported Japanese equipment, equipment designed and made in Korea, or a combination of both.

The Korean company *Pulmuone* runs a large natto facility in Goesan County, North Chungcheong. In a 2018 interview, the operators noted that their processing throughput was 300,000 packs per day, with a comfortable 81% of Korean natto market share. Notably, the owners mentioned that they started in the mid-2000s when the only options were frozen natto imports from Japan. They also noted that there were factors like climactic conditions that made natto fermentation different in Korea versus Japan.[36] *Pulmuone* focuses on small-grain black soybean natto, and has adopted sauce offerings to suit local tastes.

Korean conglomerate *CJ CheilJedang* also produces a natto product, while several companies like *JeollaNamdo* manufacture a dried *Cheonggukjang* product (a Korean fermented soybean product similar to natto, see Chapter VII). A handful of large natto manufacturers are also found in China and parts of Southeast Asia.

Most manufacturers outside of Asia have relatively small throughputs, and sell their product in commonly available food industry containers like paper and plastic take-out containers, or even glass jars, which need to be filled after fermentation. Most of these manufacturers rely on domestically available pressure cookers, custom-made fermentation rooms, and fill containers by hand.

Economy of Scale

For small scale natto producers just starting out, it's difficult to justify a natto facility larger than 30kg dry soybeans per day, especially outside of Japan where demand is relatively low. Due to their small scale, natto produced by small factories will remain expensive compared to their Japanese counterparts, and even compared to frozen, imported natto.

The price of natto is dominated by operating costs, including the cost of labour. Thus, efficient automation is key to cost-competitive natto. As discussed in the introduction to this book, without automation, non-Japanese natto will remain expensive.

If natto demand increases in a country, and there's an economic justification for constructing a large natto factory, the process might follow in the footsteps of large scale *shoyu* and *miso* factories. In these cases, Japanese companies would design and operate overseas facilities, hiring and training local staff. Examples of this in the United States are the *Kikkoman Foods, Inc.* shoyu plant in Walworth, Wisconsin (built 1973) and the *Marukome USA, Inc.* miso plant in Irvine, California (built 2002).

For example, in Thailand, *Komatsu Foods* operates a local factory in Bangkok, and uses Japanese equipment and PSP containers.

However, it's possible that non-Japanese producers end up developing and manufacturing their own natto equipment, especially if different types of packaging are developed. Packaging equipment is by far the most complicated piece of equipment in a natto production line, whereas pressure steaming cans and incubation rooms are relatively simple.

2.2 Layouts at Different Production Scales

Separation of Areas by Cleanliness

The layout of a natto facility should have an efficient and logical flow of material, and allow for separation of different areas with different cleanliness requirements.[37–39] The main process areas are separated out as follows.

Raw Material Handling: the area for handling bags or silos of dried soybeans should be physically partitioned from the rest of the factory. Raw soybeans contain the most contaminants, including dust, which is the easiest way for microorganisms and bacteriophages to get into the production areas. There is typically no direct path for operators to go from the raw material handling area to the rest of the factory without changing clothes, cleaning, and using an air shower.

Washing and Soaking: in many facility layouts, the soybean washing machine represents the boundary between the raw material handling area and the washing and soaking area. The addition of water to the soybeans captures dust, and all downstream processes can be kept dust-free. In this area, bacteria are still present, but the majority are removed during the washing process, which is crucial to prevent the beans from becoming overly acidic during soaking by the action of lactic acid bacteria. Raw soybeans can be moved to the soaking tanks by pumping with water, since they are relatively durable. Clean-in-place (CIP) systems are available for washing, soaking, and all downstream equipment.

Steaming and Inoculation: the pressure cooker is kept separate from the washing and soaking area, and acts as a gatekeeper to the rest of the natto process. During pressure cooking, sustained pressures reach as high as 2.0kg/cm^2 and temperatures as high as 134°C. This is essentially an autoclave, and practically no microorganism has a chance of surviving, including bacterial endospores. Since inoculation also takes place right at the steamer or the unloading hopper, the area should be kept separated from the washing and soaking areas. Inoculation is performed at 70–90°C to reduce the likelihood of airborne contamination.

Filling, Incubation, Maturation: the filling area requires the highest degree of cleanliness. Freshly steamed soybeans can be colonized by practically any fungi or bacteria, and the entire filling process is performed with steamed beans exposed to the factory air. Natto factories are not completely sterile, but ambient microbial counts must be low to avoid contamination. Once each package of beans is covered with a film, the chances of further airborne contamination are low. Once *B. subtilis* colonizes the fermenting soybeans, most other microorganisms will struggle to gain a foothold. A relatively clean environment must be maintained all the way until natto is ready to be loaded onto trucks for delivery.

Flow of Material

For larger facilities, multiple batches are processed in a semi-continuous manner. Thus, the layout of the factory should facilitate the flow of material in one direction. The use of pumps and gravity is important for large factories and should be taken advantage of wherever possible. Pumps, pipes, and hoses are a cheap and efficient means of transporting raw soybeans in water. The best practice is that after washing, the beans are pumped to the highest floor for soaking. They can then be dropped by gravity into individual steamers, and then dropped again by gravity from the steamer to the hopper of the filling line. In factories without multiple floors, lifting devices must be used, for example to transport steamed, inoculated beans into the hopper of the filling machine.

Separation of Production in Time

Despite efforts to keep areas of the factory separate, many small factories simply don't have the space for compartmentalized production. At a bare minimum, the raw material handling and washing area should still be separated from the rest of the facility. For the rest of the process, a small factory can maintain clean conditions by separating each process in time.

Most small natto facilities will run only one batch a day: soaked soybeans from the day prior are loaded into the pressure cooker, cooked, inoculated, and then packages are filled and loaded into the fermentation room. Once the beans are in the fermentation room, the next day's beans can be washed and soaked. Since production steps are not happening in parallel, the likelihood of cross contamination is limited. Critically, the filling process is not conducted at the same time soybeans are being washed and soaked.

Example Floorplans

Sample factory layouts are available in Sugio Watanabe's handbook on natto production.[39] Mr. Watanabe worked for Suzuyo Kogyo in the 1970s and was the director of the company's planning office. Sample layouts can also be found on Suzuyo Kogyo's website.[40]

Natto facility throughput is typically defined in units of 'bales', or kg of dry soybean equivalents, per day. Recall that one bale (*hyo*) is a Japanese unit of measurement that equates to 60kg of dried soybeans.

Less than ½ Bale (30kg) per day

This is around the smallest economically viable production scale and can be operated by a single person performing all tasks, although container filling benefits from having additional staff or some semi-automatic method of filling. No special equipment is required for washing and soaking, which is done manually in plastic buckets.

The pressure steamer, boiler, incubation room and refrigerator are required as the bare minimum equipment.

At this scale, a half-bale (30kg dry soybean equivalent) capacity steaming can is used. According to Watanabe, this size of steamer is often designed to operate at only 1kg/cm^2, which allows it to fall under a less stringent pressure class by Japanese Labour laws and standards. Internationally, some jurisdictions have a similar regulation classifying 'low pressure' as 1.02kg/cm^2 (100kPag) gauge or less (refer to your local regulations and codes). In either case, the steaming time is extended to account for the lower pressure and temperature.

The total required floor area only measures around 35m^2, with areas for washing, soaking, and filling containers, a small 2–3m^2 fermentation room, and a similar size walk-in refrigerator.

2–6 Bale (120kg–360kg) per day

The next production scale will include dedicated machines for bean washing, container filling, and final packaging.

This size of facility will have a 2–bale (120kg dry soybean equivalent) steamer, likely with automatic controls. There are still some manual operations involved. After the natto packs are filled, they are sent to a turntable where operators must pick them up by hand to load them into containers, then manually wheel the containers in and out of the incubation room and refrigerator.

At this scale, it's still common for everything to take place on a single floor. Machines are required to lift steamed beans into the hopper of the filling machine. Two or more fermentation rooms can be used, allowing multiple 120kg batches to be run each production day. A factory of this scale would require around 300m^2 total area, include storage areas and staff facilities.

6–12 Bale (360–720kg) per day

At progressively larger scales, more factory automation is added. Namely, the natto packs are now loaded into containers and stacked by machine, removing even more labour requirements. Stacks of containers are still wheeled in and out of the fermentation room and refrigerators by hand. An additional machine unloads the natto packs from their containers for secondary packaging.

At this production scale and beyond, the size of pressure cookers, filling machines, and fermentation rooms is maximized. Factories we simply operate multiple cookers, multiple filling lines, and multiple fermentation rooms in parallel. Two-floor layouts are also common at this scale, allowing soaking to take place on upper floors. A factory of this scale would require over 600m^2 total floor area.

Large Scale Facilities

Sample layouts for extremely large factories are not available in the public domain. Based on photos and videos of natto facility tours, the major features of the largest natto factories are: full automation, including robotic dollies to transport trays of natto in and out of rooms; the use of multiple (up to three) floors, with soaking on the 3rd floor, steaming on the 2nd floor; and the ability to process their own soybeans in bulk with on-site silos.

3. Food Safety

Food safety is of utmost importance in a food processing facility, and natto facilities are especially stringent. Cooked soybeans will harbour any microorganism that can gain a footing on its surface, especially since they're incubated between 38–50°C, right in the middle of the 'danger zone' of bacterial growth.

Anyone looking to start a natto business is going to have to deal with food safety regulations in the jurisdiction they're operating and selling in. A regulator unfamiliar with natto making might be skeptical or have an otherwise poor understanding of the process. This section will provide some of the necessary information needed by regulators and will help you navigate concepts involved in food safety.

This is not a food sanitation textbook, but I'll provide an abridged version of the fundamentals of food safety as they pertain to natto, along with some further reading at the end of this chapter, for those keen on learning more.

3.1 Hazards in Food

Food hazards come in three different forms:[41]

1. **Physical Hazards** are things such as metals, pebbles, and glass.
2. **Chemical Hazards** include substances such as pesticides and herbicides, cleaning chemicals, and machine oils. Sometimes **Allergens** are put in this category.
3. **Biological Hazards** include microorganisms harmful to the human body.

Physical and chemical hazards in natto-making are mostly removed at the raw material handling process. The screening process gets rid of most foreign material, while the washing process gets rid of any pesticides, dust, grease, and heavy pebbles. A metal detector or X-ray can perform a final detection on the natto packs right before incubation, immediately after filling.

By comparison, biological hazards are complex, and most of the natto manufacturer's attention is directed to mitigating these hazards.

Deadly Microorganisms

While most microorganisms are harmless, there's a handful of foodborne pathogens that are known to cause harm. The majority of these are bacteria, and they typically make you sick in the following ways.[42]

1. **Infection:** you ingest the microorganism, and they multiply in your body.
2. **Intoxication:** bacteria multiply on a food source, sometimes sporulating and/or lysing (breaking down), and release toxins into the food which you ingest.
3. **Toxicoinfection**: you ingest the microorganism, and they produce toxins within your body.

The ability of these microorganisms to propagate depends on a set of **extrinsic factors**, such as temperature, oxygen, relative humidity; and **intrinsic factors**, such as water activity, pH, oxidation-reduction potential, and nutrients.[41] Since the objective of natto making is to facilitate bacterial growth, the conditions inside a package of natto are ideal conditions for many bacteria, especially aerobic ones. Conditions are simultaneously warm, oxygen rich, high in relative humidity, high in water activity, neutral in pH, and rich in nutrients; all of which are amenable to bacterial growth.

Thus, the only measure to prevent against unwanted microorganisms is to limit their ability to contaminate a package of natto.

Transfer of Contamination

For a harmful microorganism, or 'agent' to contaminate a package of natto, and colonize it to the extent that it can cause a foodborne illness, a chain of events must occur.[43]

1. The agent must be in the environment where the food is produced.
2. There needs to be a source (reservoir) of the agent.
3. Transmission of the agent from the source to the food occurs
4. The food supports the growth of the agent (intrinsic factors).
5. The environment supports the growth of the agent (extrinsic factors).
6. A susceptible host consumes the contaminated food.

This sequence of events is borrowed from the Centers for Disease Control and Prevention (CDC) epidemiologic triangle, and CDC chain of infection. In the chain of infection for a foodborne illness, the mode of transmission is usually through the proliferation of the agent on a foodstuff, and the 'portal of entry' is usually the consumption of that food. If any one of these links in the chain of events is broken, a foodborne illness won't occur.

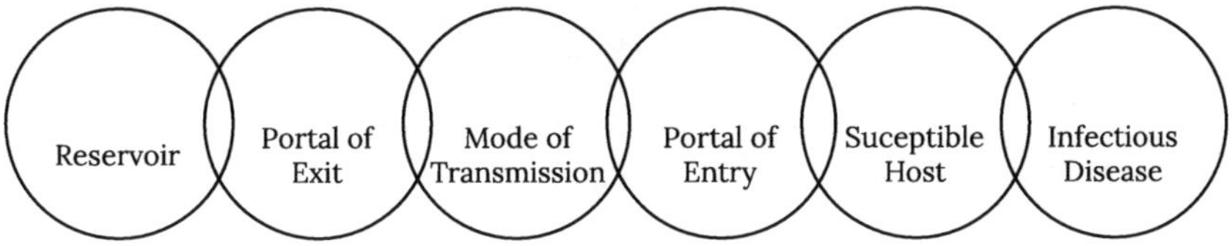

Figure II.11 CDC chain of infection

For example, you could hypothetically have the most unsanitary, unregulated natto production process, but if the host has a strong immune system against the agent (they are *not* a susceptible host), a foodborne illness will not occur. Obviously, this is not recommended; the host's immune system is the absolute last line of defense.

In natto-making, we have no control over intrinsic factors (the food, natto, supports the growth of bacteria) and extrinsic factors (the environment, i.e. the fermentation chamber, likewise supports the growth of bacteria) during the fermentation process. Factors that we *can* control are the reservoir (removing or mitigating the source of the agent), and the 'portal of exit' (mitigating the transmission of any agents from the reservoir to the food).

The Reservoir

Sources of contamination, known as the reservoir, can come from anywhere. Most commonly, these include soil, sewage, live animals (including pests and insects), external surfaces, the ingredients themselves, and workers.[37,38] Our goal is the stop these reservoirs from entering packages of natto and propagating, which is the main objective of sanitation in a natto facility.

List of Agents and Reservoirs Encountered in Natto

A list of microorganisms that are problematic and commonly found in the natto-making process has been developed by the Japan Natto Cooperative Society Federation.[44] This list also contains the effects of each problematic microorganism on the resultant natto, common reservoirs, and methods of removal.

In addition to preventing contamination by harmful bacteria, producers also need to worry about microorganisms and viruses that will harm or compete with *Bacillus subtilis,* compromising the final product quality, even if this doesn't lead to foodborne illness. This includes other members of the genus *Bacillus*, and importantly, the bacteriophage of *Bacillus subtilis*. This bacteriophage is a virus that targets only *B. subtilis*, and while harmless to humans, it will happily destroy a colony of *B. subtilis*, compromising the natto's quality and leaving it susceptible to further microbial contamination.

Table II.7 Commonly problematic microorganisms and viruses in natto making.[44]

Microorganism	Effects on Natto	Reservoir	Removal & Notes
Bacteriophages			
Bacillus subtilis **bacteriophage Type I–IV**	The finished natto is the same as steamed soybeans Weak threads when stirring	1) Soil, dust 2) Drains, floors, ceilings, etc.	Heat: 65°C for >10mins Chemicals: Alcohol, sodium hypochlorite, invert soap, etc. 1) If contamination is significant, inhibits the growth of Bacillus subtilis
Aerobic spore-forming bacteria			
Bacillus licheniformis	Appearance remains the same, but threads and aroma are weak	1) Soil, dead grass, rice straw 2) Airborne as spores 3) Drains, floors, ceilings, etc.	Heat: 121°C for >15mins 1) Difficult to kill due to presence of spores 2) Same family as Bacillus subtilis var. natto but competes during fermentation 3) Difficult to test for because they are homologous
Bacillus subtilis (non-natto)	Slightly reddish appearance, and weak or lumpy threads		
Bacillus pumilus	Bacterial cover is good, but flavour is lost		
Bacillus cereus	Slightly red, with unpleasant odour		
Bacillus coagulans	Lumpy threads, and poor taste		
Cytobacillus firmus	Lumpy threads, and poor taste		
Anaerobic spore-forming bacteria			
Clostridium spp.	Causes on offensive, sweat-like odour, caused by lack of acidity	1) Soil, water, intestinal tracts of animals and humans; airborne as spores	Heat: 121°C for >15mins 1) Difficult to kill due to presence of spores
Anaerobic non-spore forming bacteria			
Lactobacillus spp.	Lactic acid production during soaking lowers the pH of the soybeans, preventing bacillus from propagating	1) Airborne, reproduces in the lower part of soaking tank	Heat: varies 1) When soaking is completed, thoroughly clean soaking tank with weak alkaline chemical cleaning agent

Filamentous fungi			
Rhizopus spp. ***Penicillium spp.***	Filamentous fungi rarely affect natto itself. Spores may survive in the final product	1) Soil, dust, ceilings, and walls	Heat: 80°C for >10 mins Chemicals: Sodium hypochlorite, soap

This list of viruses and microorganisms is not exhaustive, and other microorganisms should be considered when identifying biological hazards in the natto making process.

Destroying Microorganisms

Methods to destroy or control microorganisms include heat, chemicals, and radiation.[42]

Heat is by far the most important. One major step of natto-making is the pressure steaming process, where soybeans are subjected to temperatures of 121°C–134°C (at pressures of 1.0–2.0kg/cm^2) for 40–60 minutes. The pressure cooker acts as an autoclave and practically no microorganism survives this step.

Chemical destruction methods form the basis of sanitation and cleaning. Chemicals, which include bleaches, soaps, and sprays, are used to clean people, equipment, and the facility.

Radiation is seldom used in the natto-making process, except in some cases to the irradiate packaging material prior to shipment. This is done by the packaging manufacturer.

Controlling Microorganisms

Even if a few problematic bacteria make their way into a sample of food, there are other ways of mitigating their growth. Common methods include increasing the osmotic pressure, dehydration, biopreservation and refrigeration.[42] The latter two are relevant to natto making.

Biopreservation is the use of biological agents to control unwanted bacterial growth. The mechanism relevant to natto is called blanketing, whereby the growing culture of *Bacillus subtilis* creates a robust biofilm and is given a numerical advantage during inoculation, preventing other bacteria from adhering and propagating.

Refrigeration is the use of low temperatures to mitigate the growth of any microorganism, and is maintained from the moment natto is pulled out of the incubator, all the way to point of consumption.

The Need for a Food Safety Program

In summary, there are many factors involved in food safety. Standing between a harmful agent and a susceptible host, the goal of a food processor is to diligently maintain layer upon layer of safeguards to prevent the occurrence of a foodborne illness and keep public health and safety as paramount. Therefore, we need good attention to detail with multiple programs and procedures that form the foundation of a functioning processing plant.

3.2 Prerequisite Programs

The process of eliminating food hazards can be thought of as a series of protective layers of defense that prevent any unwanted physical, chemical, or biological hazards from reaching the consumer. These include everything from the design of the facility and equipment itself, to agreements with suppliers, cleaning procedures, and procedures for staff. Examples are provided below.[45]

1. **Facility Design:** the natto facility must be designed, constructed, and maintained according to sanitary design. This was examined in Section 2.
2. **Supplier Control and Specifications:** the suppliers of all raw materials, including beans, municipal water, natto packages, sauce, and inoculum, must meet stringent specifications and be examined periodically.
3. **Production Equipment:** all equipment must be designed to allow for proper cleaning. This is on the onus of the equipment manufacturer, and in natto's case, Suzuyo Kogyo or Miwa Seiki.
4. **Cleaning and Sanitation:** all workers must carry out Sanitary Standard Operating Procedures (SSOPs), which are the procedures to clean equipment, as opposed to Standard Operating Procedures (SOPs) which relate to the natto production itself.

5. **Personal Hygiene:** all workers must be trained in personal hygiene, which includes not working when sick or harbouring open sores, and being clean and neatly groomed.
6. **Chemical Control:** cleaning chemicals, and chemicals used for maintenance such as lubricating oils and lab chemicals, must be stored away from the food production environment.
7. **Receiving, Storage, Shipping:** storage and shipping of finished natto products must be kept within a cold chain.
8. **Pest Control:** pests such as insects and rodents must be controlled through facility design features, traps, or otherwise.
9. **Traceability and Recall:** if something does go wrong, there needs to be a way to trace the problematic batch of natto, and initiate a recall based on the lot number.
10. **Training:** workers must be trained and refreshed on all the programs listed above.

The complexity of each program depends on the nature of the facility. These procedures are often called Pre-Requisite Programs (PRPs), and they form the basis of other food safety programs such as Hazard Analysis and Critical Control Points (HACCP). Creating a complete HACCP plan is beyond the scope of this book, but the information provided in this chapter can be used for some steps of the HACCP plan, such as developing a flow diagram and conducting the hazard analysis.

The Natto-Ren HACCP Guide

For industry expertise on controlling microbial hazards, we can also look towards the HACCP guide published by the Japan Natto Cooperative Society Federation (Natto-Ren) from March 2008.[44]

This document, which is available to the public, is geared towards small natto factories with less than 50 staff, and provides examples of a process flow diagram, process descriptions, a sanitation management plan, a hygiene management plan, and templates for record-keeping.

4. Homemade & Lab Scale Natto

I want to dedicate a final section to natto making at small scales, either in the home kitchen or in research environments. Although natto making is a relatively simple process, I encourage anyone who hasn't tried natto to firstly try a commercial product, in order to understand its flavour and aroma, because natto is hard to describe. This will allow you to properly assess the final product when you make it at home.

Japanese and Korean grocery stores abroad are likely to stock natto, along with small size health food stores that work with local independent producers. Certain Chinese and Southeast Asian grocers also stock natto.

The process for homemade natto is similar to the one described in Section 2, but I'll provide some tips that apply to homemade or lab scale natto only. Additional procedures for lab scale natto, used for new product development, are detailed in Chapter IV.6.

Sourcing Natto Beans

The most important starting point for making natto at home is to source the correct type of soybeans. Normal food grade soybeans that are typically found in Asian grocers or bulk food stores are larger than those used for natto. If you make natto with these types of beans, often the resulting natto will be hard, not sufficiently stringy, and retain a beany taste and aroma.

Natto grade soybeans are less common in grocery stores, but in general we are looking for small grain soybeans specifically bred for natto making. Both Canada and the United States have agricultural divisions dedicated to growing soybeans for the natto market, but almost all of these soybeans are exported to Japan. Natto grade soybeans can also be found from specialty retailers, and soybeans destined for sprouting are also usually small grade.

If small grain soybeans are difficult to obtain, I suggest extending the total fermentation time from 20–22h to 24h or more for large soybeans. Trial-and-error may be needed to dial in the correct fermentation time and temperature. Large soybeans can also be split to make *hikiwari natto* using a corona mill or otherwise, and can be dried in an oven before splitting.

Washing and Soaking

Washing and Soaking proceeds identically to industrially produced natto. Wash thoroughly to remove dust and debris, and discard the water. Refer to Section 1.1 for details on soaking time and temperature. The standard laboratory method for natto is to soak at 20°C for 16h.

Pressure Steaming

Pressure steaming is recommended for cooking the soaked soybeans. Household pressure cookers are readily available in either electric or stovetop forms. These pressure cookers typically operate at between 0.87–1.02kg/cm^2 (85kPa–100kPa) as a safety requirement. Fill the pressure cooker with around 3cm of water and suspend the soaked beans in a colander or basket, above the water level. With a pressure cooker, steam for 40 minutes, starting when the cooker reaches pressure. Allow the cooker to depressurize naturally for 10 minutes before venting and opening.

An alternative is to steam or boil at atmospheric pressure. However, this is energy intensive at will take between 3–5 hours of constant simmering or steaming, and is not recommended.

The standard laboratory method is to cook the soybeans in an autoclave, in which soybeans are steamed at 131–133°C for 30 minutes.

As described before, the soybeans should be steamed until soft enough to crush easily between the index and ring or pinky finger.

Sanitization

All tools that come into contact with the steamed beans and spore solution should be sanitized with a 50–200ppm bleach solution, 70% isopropyl alcohol, a mild alkaline detergent, or equivalent; then rinsed with boiling water.

These include the natto container, mixing containers (if used), any spoons or mixing utensils, and thermometers.

Inoculation

Inoculate right after the beans are removed from the pressure cooker, between 70–90°C, with a suspension of spores dissolved in boiled, then cooled water, or dissolve the spores in the hot broth left behind after steaming, at the same inoculation rate described in Section 1.2, or as suggested by the spore manufacturer. Mix the spore suspension and steamed beans thoroughly after inoculation, without damaging the soft beans.

Spores can be found from several online retailers. Takahashi and Miyagino both provide spore suspensions for home use.

An alternative is to inoculate with an existing package of natto. However, this should be done at temperatures below 50°C since we are using mostly vegetative (i.e. non-sporulated) *Bacillus subtilis* cells. This increases the chance for contamination while the beans cool, and increases the total preparation time.

For those seeking a challenge, spore suspensions can be made with some basic wet lab equipment and microbial techniques. This may be the only way to acquire 'specialty' spores such as non-sticky and low odour natto, and the techniques are described in Chapter IV.

Incubation

Incubate at 38–40°C in accordance with Section 1.3. Homemade incubation setups hold much lower total mass than industrial setups, so temperatures can rise and fall more rapidly. Industrial natto producers will start cooling at 16h, but the natto will only reach 39°C at 20h. For home producers, the incubator should be kept at 39°C for 20h to replicate the same conditions. Longer fermentation periods can be used for bigger beans.

The standard laboratory method is to incubate the soybeans at 39°C, for 18h at 90% RH, then drop the temperature to 20°C for 2h at 50% RH. This is done in specialized, programmable incubators with standard PSP packs.

An incubation method with relatively accurate temperature control is desirable: options include dehydrators, bread proofers, yogurt makers (including those provided as a function with some electric pressure cookers), and jerry-rigged setups with a temperature controller and heating element. If you have an accurate kitchen thermometer or probe, consider how the actual temperature profile of the incubator behaves relative to its programmed setpoint and deadband (the band of temperature values in which an on-off controller will do nothing). For example, temperature controllers with a setpoint of 39°C might only activate the heating element at 35°C, and switch it off at 39°C, leading to an average temperature of 37°C. These minute differences can change the required fermentation time.

It's important to keep a humid microenvironment within the natto during fermentation, but still allow for air exchanges to replenish oxygen consumed during the fermentation process. This is usually achieved by applying a plastic film on top of the natto, which can be made with cling film, parchment paper, wax paper, or aluminum foil (with pinholes added at around 1 per $4cm^2$).

For incubators with a relatively large amount of air exchanges, such as dehydrators, a moist environment can be kept by covering the natto with two separate layers of cling film, with small holes poked on the surface at 1–2cm intervals.

Cooling and Consumption

Remove the natto from the incubation environment and allow to cool in the refrigerator for a minimum of 24h before consuming. The standard laboratory method is to allow samples to mature at 5°C for 24h before assessing the quality.

Summary

Natto production is not a complex process, but attention to detail is required for all steps of production to make consistent and high quality natto. Production equipment is readily available in Japan, but for overseas producers, making cost-competitive natto will remain a challenge without the help of highly automated equipment. Lastly, food safety is paramount for the natto-making process since the incubation process is performed right within the 'danger zone' of bacterial growth.

In the next chapter, we'll take a look at how natto is eaten, the standards for properly assessing its quality, and some of the market and industry trends of natto in Japan.

Further Reading

Sugio Watanabe and Kan Kiuchi, among a handful of other authors, are by far the most prolific when it comes to writing about the natto production process. Their writing can be found in the English language source:

- **Industrialization of Japanese Natto** (in Industrialization of Indigenous Fermented Foods).[20]

Sugio Watanabe, who has worked for Suzuyo Kogyo Co. Ltd., has also published two handbooks dedicated to natto production. Although these are in Japanese, they were translated to support the contents of this Chapter:

- 納豆入門 [**Introduction to Natto**], published 2020.[46]
- 納豆:原料大豆の選び方から販売戦略まで [**Natto: From how to select raw soybeans to sales strategies**], published 2002.[47]

Photos of natto production equipment can be found on Suzuyo Kogyo's website:

- **suzuyokogyo.com**

Videos of natto production can also be found online by searching for 'Natto Factory' in Japanese [納豆工場]. As of writing, two major producers, *Tokiwa Shokuhin* [登喜和食品] and *Kasanago Shokuhin* [金砂郷食品] have YouTube channels depicting the entire natto making process with modern equipment.

III. Natto Consumption

With only three companies producing *Bacillus subtilis* spores (*natto-kin*) for commercial sale, and only one or two manufacturers for natto-making equipment, natto products are somewhat uniform across the industry.

Variety in natto boils down to variations in bean size and cultivar, and different types of sauce. Some producers have also bred unique strains of *B. subtilis* that produce a low odour natto, a low stickiness natto, and others, with varying degrees of commercial success. As of writing, *natto-kin* of these strains are not available for commercial sale.

Smaller volume producers, or the development arm of larger companies, have a bit more flexibility when it comes to product development. To differentiate themselves from the ocean of mass-produced natto, they will incorporate the use of different strains and types of soybeans (including brown soybeans, black soybeans, edamame, etc.), and mix in unique ingredients such as buckwheat, dried daikon, barley, and even yuzu peels.

Per-capita natto consumption in Japan remains strong, and has been on an uptick between 2011 to 2019, as facilities recovered from the Great Recession of 2008 and the 2011 Tohoku Earthquake.[48] In recent years, a renewed interest in fermented foods, coupled with historically strong marketing on natto's health benefits, has helped maintain a stable market in Japan.

Natto producers are still keen to open new markets, especially overseas and among people who generally dislike natto. There remains the challenge of getting new customers past natto's peculiar taste and texture. For those who are squeamish, there are many ways to eat natto, and different types of natto, that are more amenable to first timers.

In this chapter, we'll look at how natto is consumed, the variety in natto offerings, products made of natto, market data, future developments, and how to properly evaluate its quality.

1. Eating Natto and Product Diversity

1.1 Natto Basics

Packaging

Figure III.1 Typical features of a natto label (above label is fictitious).
Note: producers often choose to highlight the origin, soybean size, or other features.
Exported natto includes a lot number only, and no best before date.
See Glossary of Japanese terms for commonly used label text.

Most natto comes in a 50g serving square polystyrene paper (PSP) pack, which, when opened, usually contains packets of sauce (*tare*) and mustard (*karashi*), which are the most common accompaniments, and an additional piece of film atop the natto. Some producers have the sauce contained inside a cavity in the lid, which is cracked open to let out the sauce, while other packages have no sauce at all (*tare-nashi*).

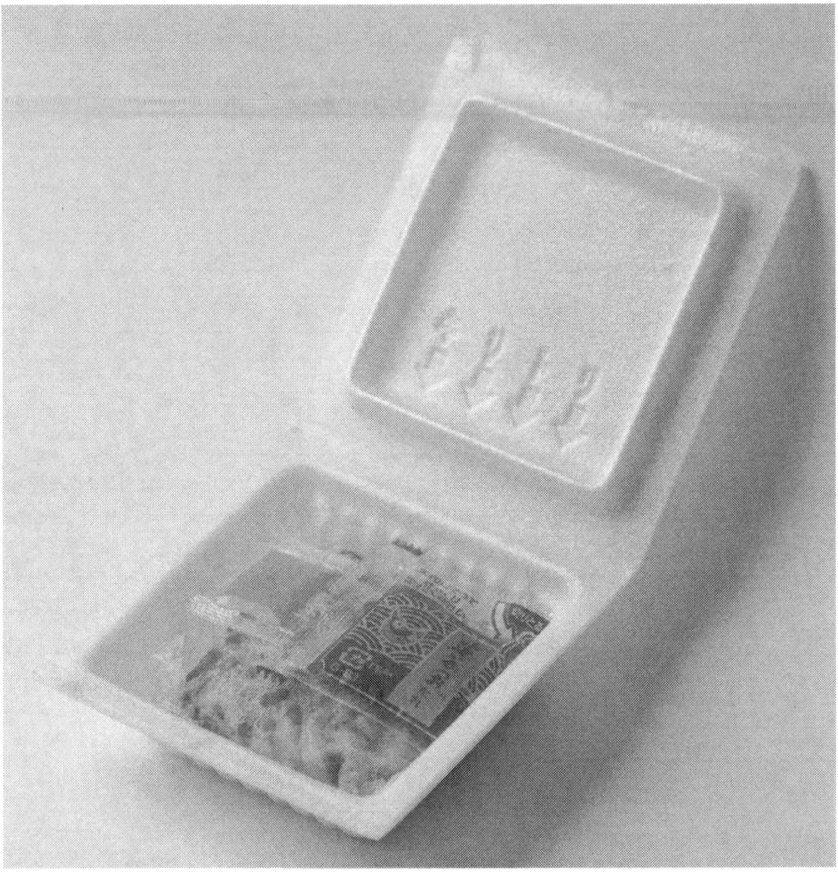

Photo III.1 Opened 50g PSP pack of natto containing tare and karashi

The second most common form of natto packaging is in 30g paper or plastic cups, followed by less common packaging materials such as thin sheets of pine veneer (*kyogi*), bamboo mats (*sudare*) and traditional rice straw bundles (*wara* or *tsuto*), whose physical properties were described in Chapter II. Plant-based packaging types are usually reserved for 'craft' natto that is trying to distinguish itself among the commodity offerings.

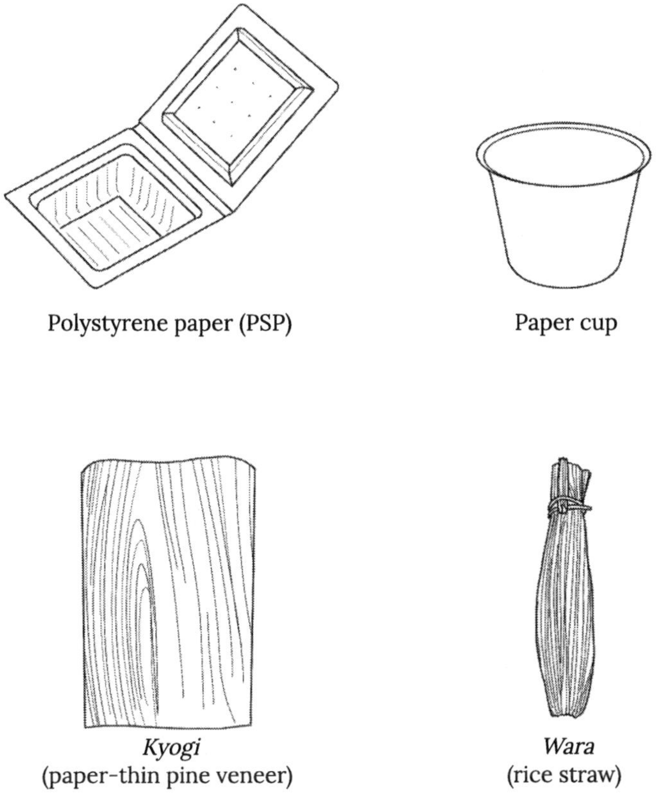

Figure III.2 Natto packaging examples

Within the natto section of Japanese supermarkets, producers will opt for bold, colourful packaging to compete for the consumer's attention, and there are about as many natto products as there are combinations of soybean types, and sauces to pair them with. Packages will identify the following:

1. The manufacturer, and any special branding.
2. Soybean type, origin (especially if Japanese), and size.
3. Name of the product, sometimes with flavour text.
4. List and description of accompaniments, if any.
5. Package weight, expiry, and note to keep refrigerated.

Photo III.2 Left: three tier PSP pack of natto exported for North American sale
Right: individual PSP pack

Stirring Natto

The 'correct' way to eat natto is to take a pair of chopsticks and stir vigorously until the biofilm coating the beans whips up into a frothy, white substance. Poly-γ-glutamic acid and levan, the two major components of the biofilm, are transparent, so the white colour is caused by the incorporation of air into the mucilage.

After stirring, sauce is added as desired, and the whole pack is stirred again to incorporate the sauce, then emptied out: usually onto a bowl of rice.

Anecdotally, people will claim that stirring the beans before adding sauce 'brings out the umami' of the natto. While stirring can have several affects, mechanical agitation won't affect the concentration of any *umami* components. Amino acids, nucleic acids, and the action of enzymes would not be significantly affected by a brief stirring period.

Stirring might improve both the perceived mouthfeel and aroma of the natto. In the same way *ramen* is slurped, or how wine testers slurp and aerate a mouthful of wine, the addition of air helps gaseous aromas travel up the back of the nose. The increased volume of the natto mucilage also thickens the sauce, allowing it to linger on the tongue and create a fuller mouthfeel.

Many people dislike this slimy texture, known by its Japanese onomatopoeia *neba-neba*, which is also found in foods like okra and grated

Japanese mountain yam. For those consuming natto for the first time, or for those who simply don't like the gooiness, stirring is optional. Alternatively, add the sauce to the natto first, and then stir lightly to incorporate.

The Taste of Natto

Natto itself is relatively neutral, and may have a strong aroma. This might lead one to ask, why eat natto anyways? Why not just eat boiled or steamed soybeans directly, and save the hassle?

To answer this question, I encourage you to try the following experiment. Make natto as usual, but reserve some of the steamed soybeans and refrigerate them without fermenting. Then, eat the steamed soybeans, with a side-by-side comparison to natto, both with the same type and amount of seasoning.

It's difficult to describe how unpalatable plain, steamed, or boiled soybeans are. They aren't bitter, sour, pungent, or immediately offensive; and yet they still taste unpleasant. Steamed soybeans have this odd beany odour that I can only describe as a mix of wet cardboard, grass, and paint.

By comparison, natto is completely different and takes on a unique bouquet of aromas that is a mixture of nuttiness, earthiness, cheese, and fruity esters. People consuming natto for the first time have used cheese and coffee as tasting notes, but the exact combination of flavours is unique to natto and difficult to describe.

The Role of Sauce

Putting aside all the strange olfactory qualities of natto, the taste of natto is heavily dependent on the sauce used. This sauce mixes with the highly sticky natto and loosens the texture.

Both the sauce and mustard compliment the taste of the natto. Natto is normally alkaline and not sweet, so the addition of sauce and mustard gives the natto some much needed sweetness and acidity. Mustard also contains various isothiocyanate compounds that can mask or compliment the odour of natto.

Natto Combinations

Besides sauce and mustard, many other toppings are popular with natto. When asked, "What do you put in natto? (Please select all that apply)" as part of Natto-Ren's biyearly survey, the most popular additions and toppings in 2021 surveyed were as follows.[49]

Table III.1 Natto accompaniments by popularity (2021)

Selected (%)	English	Japanese
84.7	Accompanying sauce	添付のタレ
52.5	Mustard	からし
37.5	Green onion	ネギ
18.6	Soy sauce	しょうゆ
17.4	Kimchi	キムチ
15.3	Eggs, whole	たまご（全卵）
7.8	Nori	海苔
6.4	Mentsuyu (noodle sauce)	めんつゆ
6.3	Okra	オクラ
5.4	Eggs, yolk	たまご（黄味）
5.2	Oroshi (grated daikon radish)	大根おろし
4.3	Shirako / shirasu (whitebait)	しらす（じゃこ）
4.2	Sesame seeds	ごま
3.9	Katsuobushi (bonito flakes)	かつおぶし
3.6	Vinegar	酢
3.5	Shiso	しそ
3.2	Umeboshi (pickled plum)	梅干し
3.0	Yamaimo (mountain yam)	山芋
2.9	Tofu	豆腐
2.8	Aonori (type of dried seaweed)	青のり
2.6	Mayonnaise	マヨネーズ
2.5	Tororo (grated yam)	とろろ
2.4	Wasabi	わさび
2.4	Olive Oil	オリーブオイル
2.3	Chili Oil	ラー油
2.1	Myoga (Japanese ginger)	みょうが
2.1	Nametake (Mushroom)	なめたけ
1.6	Chopped up tuna	マグロのブツ切り
1.5	Pickled Ginger	しょうが
1.4	Chopped up squid	イカのぶつ切り
1.3	Togarashi (Red chilli pepper)	唐辛子
1.2	Turnip Greens	野沢菜

Selected (%)	English	Japanese
1.1	Sugar	砂糖
0.9	Konbu	昆布
4.9	Others	その他
3.0	Nothing	何も入れない

Notably, other foods that have the *neba-neba* texture (okra, yamaimo, tororo), as well as acidic items, fresh herbs, and seaweeds, all seem to be popular items.

1.2 Product Offerings

Sauce and Mustard

Around 80% of natto is sold with packets of sauce (*tare*). Over the years, the amount and salinity of this sauce has become relatively consistent. In a survey conducted of 11 natto brands, most natto packages use sauce with around 13% salinity, which mixes into the natto for around 1.1% salt when ingested, while reducing the pH to an average of 6.2.[26]

Most sauce offerings include ingredients typical of *tare*, and are based around *shoyu* as a base. The addition of sugars, acids, and vinegar are commonplace in sauce mixtures. Producers sometimes try to differentiate themselves by using sauces with familiar flavour profiles, such as *umeboshi, shiso, oroshi*, egg yolk *shoyu*, butter *shoyu*, garlic *shoyu*, *yakiniku* sauce, and *unagi* sauce, just to name a few.

There is nothing particularly unique about the mustard used in natto. Like with natto grade soybeans, Canada is a major supplier of mustard seed to Japan. The mustard used with natto is generally a 'hot mustard' or 'brown mustard', made with the black and brown seeds of *Brassica nigra*, or 'oriental mustard,' made with the yellow seeds of *Brassica juncea*. 'White mustard', made with the seeds of *Sinapis alba,* are rarely used.[26] Mustard is made by grinding mustard seeds with water, which causes the enzyme myrosinase to hydrolyze glucosinolate, forming mustard oils such as allyl isothiocyanate that cause its pungency.

Myrosinase and glucosinolate are stored separately in mustard plant cells, and only combine when plant tissue is chewed on by an herbivore, releasing the mustard oils in a process named the 'mustard oil bomb'. This is supposed to be a deterrent, but humans like the spiciness.[50] Allyl isothiocyanate is also the cause of spiciness in most *Brassicaceae* including *wasabi* and horseradish.

In commercial preparations, ground mustard seed is mixed with salt and vinegar to preserve its flavour and shelf life. Its high oil content makes it difficult to prepare as a powder.

Table III.2 Sauce composition of 11 natto samples[26]

Sample No.	1	2	3	4	5	6	7	8	9	10	11	Avg
Sauce Composition												
Total N (%)	0.9	0.9	0.6	1.0	1.0	1.1	0.6	0.8	1.1	1.3	0.6	0.9
Salt (%)	12	12	11	12	14	14	7	14	12	13	12	12
Non-salt TDS(%)	24	24	19	27	26	29	21	37	41	19	26	27
Glut. acid (%)	2.4	3.0	2.8	4.6	0.9	1.3	2.0	4.5	4.2	1.5	1.8	2.6
Inosinic acid (%)	0.1	0.1	0.1	1.0	0.1	0.1	0.1	0.1	0.1	0.2	0.2	0.2
pH	4.5	4.3	4.5	4.7	4.1	4.4	4.5	4.2	5.0	4.5	4.8	4.5
Lactic acid (%)	0	0.5	0	0.2	0.2	0.2	0.1	0.1	0.1	0.3	0.1	0.1
Acetic acid (%)	0.1	0.1	0.2	0.1	0.3	0.2	0.3	0.8	0.1	0.2	0.7	0.3
Citric acid (%)	0	0	0	0	0	0.2	0	0.3	0.1	0	0	0.1
Properties, Sauce mixed with Natto												
Salt, natto w/ sauce (%)	1.1	1	1	1.2	1.3	0.8	1.1	1.3	1.3	0.9	1.1	1.1
Natto pH with seasoning	6.5	6.7	6.2	6.1	5.7	6.4	6.3	5.7	6.4	5.9	6.2	6.2
Seasoning, %wt. of natto	12	10	11	13	13	7	22	13	16	9	13	13

(%) All percentages in w/v (g per 100mL), N: nitrogen, TDS: total dissolved solids

Bean Varieties

For the soybean, any cultivar can be used, but with varying levels of success. Smaller soybeans, thinner seedcoats, and higher sugar content are generally desirable for natto.

However, there are several natto products that use of other types of soybeans, including heirloom varieties such as black, red, and green soybeans, as well as immature soybeans (*edamame*), of all different sizes. Producers will adjust the production process for different types of beans, relying on their sensory evaluation to determine if natto made from these beans was produced successfully. Larger and thicker skinned soybeans will require longer fermentation time, and vice versa.

Non-soybean natto is rarely attempted by commercial producers. Natto hobbyists and enthusiasts have attempted natto with different types of beans such as chickpeas, mung beans, black beans, and black-eyed peas, with varying degrees of success, and with some or no documentation. Historically, these types of beans are more palatable without fermentation, which is the reason why natto isn't usually made with beans like mung beans or adzuki beans. However, they still have value as a soybean alternative for those with soy allergies.

2. Cooking Natto and Processed Natto

2.1 Dishes Made from Natto

Natto on rice is by far the most popular method of eating natto. However, there are a handful of other dishes that can be made with natto. Notably, *natto-maki* (as a normal sushi roll, or *gunkan-maki*: battleship sushi) are common across convenience stores in and in casual sushi restaurants. Natto is also found as an ingredient for noodle dishes, pasta dishes, and on bread. Other examples are found in Natto-Ren's survey. When asked: 'What types of dishes has the consumer prepared at home? (select all that apply)', responses were as follows:[49]

Table III.3 Home cooked natto dishes by popularity, 2021

Selected (%)	English	Japanese
42.9	Natto maki (sushi, gunkan-maki)	納豆巻き（おすし）
21.2	Natto soba / udon (cold)	納豆そば・うどん
19.9	Natto fried rice	納豆チャーハン
19.3	In miso soup as natto-jiru	味噌汁に入れる（納豆汁）
16.2	Inside aburaage	油揚げに入れる
15.9	Natto spaghetti	納豆スパゲッティ
14.4	Cooked with eggs	卵料理に加える
13.5	Natto tofu (ex. cold tofu)	納豆豆腐（冷やっこなど）

Selected (%)	English	Japanese
11.6	With chopped or sliced squid	イカ納豆
11.1	On Bread (natto toast)	パンにのせる（納豆トースト）
10.9	Natto curry	納豆カレー
9.7	Tuna natto	マグロ納豆
9.2	Rice bowl (i.e. bakudan-don, etc.)	丼もの（ばくだん丼、海鮮丼など）
5.1	With vegetables dishes (aemono)	和え物
4.3	Crushed as baby food	離乳食用につぶす・きざむ
4.2	Tempura or deep fried	天ぷら・フライなど揚げ物
4.0	Natto okonomiyaki	納豆お好み焼き
3.3	Salad topping	納豆乗せサラダなど
3.0	Ramen / udon (hot)	納豆ラーメン、納豆うどんなど
2.7	Kimchi-nabe	キムチ鍋に入れる
2.7	Dumplings	納豆餃子
2.1	Stir-fry dishes	炒めもの
1.8	Pizza	納豆ピザ
1.5	Others	その他
23.5	Never tried any of the above	調理されたものを食べた事がない

2.2 Products Made from Processing Natto

There is also a small subset of the natto market that includes products made by further processing natto. These are considered niche products, and are usually found in specialty stores or alongside smaller producers.

Dried Natto

Dried natto was commonly consumed during tea ceremonies in Ibaraki and Tochigi prefectures, although nowadays it is sold in many supermarkets and roadside markets as a snack. There are several methods for producing 'dried natto', either by air drying, freeze drying, or deep frying.

Dried natto can be made by air drying, typically as *hoshi natto*, to the point where it is either hard and brittle, or slightly pliable. One method for producing *hoshi natto* given by Watanabe is as follows:[32]

1. Natto is produced in large containers and covered with a perforated film, and fermented as usual.

2. Salt is added at 3% by weight, stirred, and left with the natto for 2–3 days.
3. The salted natto is placed on the plastic container of a hot-air dryer at 2cm thickness and partially dried for 10h. Drying methods vary and include low temperature vacuum drying and freeze-drying.
4. Flour is added to the partially dried natto and mixed. The mixture is then dried further.
5. Excess flour is removed with a sieve. The *hoshi natto* is then packaged and heat sealed.

The author of the above method is light on details such as drying temperature, duration, and final moisture content. Based on available nutrition labels, *hoshi natto* appears to have a moisture content of 14%–16%.[51]

Another form of dried natto was popularized by Japanese Airlines (JAL) with the brand name *dorai natto*, which was provided as an in-air snack in the same role as a package of peanuts. JAL's product is manufactured by a company called Forza Co., Limited, based out of Tokyo, but other companies also make equivalents of *dorai natto.* Their website states that the natto is deep-fried using a method called 'decompression frying', or frying under vacuum, to reduce the boiling temperature of water and preserve some of the natto's bioactive compounds.[52] However, there have been no academic studies confirming the efficacy of this method, nor details of the process available in the public domain. *Dorai natto* resembles roasted, salted peanuts in flavour, and has a much more subdued aroma and no sticky texture. Moisture content ranges from 4–5%. The natto raw material of *dorai natto* can be fermented in bulk just like *hoshi natto.* Notably, the company Kasanago Foods, the makers of a non-sticky natto, also manufactures a version of *dorai natto* called 'Parintou'.[53] In this case, the non-sticky bulk fermented natto makes processing and deep-frying the raw natto much easier.

Lastly, dried natto can also be made by freeze-drying. This method preserves most of the natto's nascent characteristics, and it can be reconstituted simply by adding water, making it a valuable emergency preparedness food.[54] Moisture content of freeze-dried natto is between 1–2%.

Note: all moisture contents calculated above were based on label information, and assuming the ash content is 11.5% of protein content (identical to standard natto).

Yukiwari or Goto Natto

Yukiwari natto is a specialty of Yamagata Prefecture, where it is known locally as *goto natto*. It is made by preserving natto with rice *koji* (steamed rice fermented with the fungi *Aspergillus oryzae*), and salt; making it like *miso*. The basic ingredients are the same as miso, although the preparation and taste are quite different. The name *goto natto* originally referred to the use of brewing barrels to preserve the natto mixture. The ingredients are as follows.[32]

1. 13.5kg *hikiwari natto*
2. 7.5kg rice *koji*
3. 3kg salt (around 12.5% by weight)

Rice *koji* provides strong enzymes from *Aspergillus oryzae* that allow it to further break down the components of natto, especially starches and other insoluble carbohydrates. The preservation time is unclear, but according to historians it was prepared during the winter and eaten during the rice-planting season in spring.

Touzou

Touzou is a specialty of Isumi District and Ichihara City, both from Chiba prefecture, and in Ibaraki prefecture. Like *yukiwari natto*, it is a mixture of natto, *koji*, and salt, but also contains daikon and soybean broth. Preparations vary, but it was likely developed to make use of the bean broth that was a by-product of the natto-making or miso-making process. The ingredients are as follows:[32]

1. 1.6kg natto
2. 1.5kg *koji*
3. 4kg daikon
4. 1.6kg salt
5. 7kg soybean broth

The daikon is first cut, dried outdoors, then soaked with boiling water. The entire mixture is then combined and matured for around 20 days.

Soboro Natto

Soboro natto is a preparation from Ibaraki Prefecture. Natto is mixed with cut, dried daikon, seasoned with shoyu, salt, and *koji*, and served as a side dish with *sake*, or in *ochazuke*. It is similar to *touzou* in the use of dried daikon, and the two preparations may have a common origin. Originally made by individual families, it is also available in some supermarkets. The ingredients are as follows:[32]

1. 10 parts *itohiki natto*
2. 10 parts cut, dried daikon
3. 1.5 parts salt
4. 1.5 parts *shoyu*
5. 6 parts *koji*

The cut, dried daikon is soaked in hot water to soften it, prior to mixing. Then, all ingredients are mixed, and it is left to sit for 4 days in the summer or 10 days in the winter, then refrigerated before serving. Recipes vary, but the key ingredients are natto, daikon, and *shoyu*.

Sugio Watanabe's "Gingo Natto"

I want to give a mention to Sugio Watanabe's *ginjo natto*, which is not a commercial product but more than an off-topic musing within his book, 'Introduction to Natto'.

Around year 2000, Mr. Watanabe received a large case of natto from a manufacturer in early December. After sharing some with his neighbours, he kept it in the back of his fridge, and like many of us would have done, completely forgot about it until the middle of January in the new year. He opened the package, thinking it would be inedible due to tyrosine and struvite, but instead found it to look like *kintsuba* (a confectionary made with *tsubuan*, sweet red bean paste, and coated with flour), with a brown interior and a white frost-like coating.

In the true spirit of fermented food discovery, Mr. Watanabe ate it and found it had no ammonia odour and tasted rich, with a thick umami and a slightly sour flavour.

Mr. Watanabe named the product *gingo natto*, noting that shoyu and mustard were good compliments, just like regular natto. However, since it takes 30–40 days of refrigeration, so he mused that it was far from commercially feasible.[32] It is also not recommended to consume natto a full one month after its best before date.

My best guess is that natto simply dried in cool temperatures, partially concentrating its flavour, and may have been colonized by *Lactobacillus* or *Acetobacter* to produce sour lactic acid or acetic acid. Enzymatic digestion during the long storage time may have yielded additional umami, and sour amino acids.

3. Evaluating Natto

There is a standard way of evaluating a package of natto, using both qualitative and quantitative methods. This is conducted immediately after fermentation, and after primary refrigeration, and allows the natto maker to assess the quality of the product prior to shipment.

3.1 Qualitative Evaluation: Sensory Test Methods

The qualitative evaluation of natto involves conducting an organoleptic test (its flavour, odour, appearance, and mouthfeel) using your own senses, and without any analytical instruments. When conducting a qualitative evaluation, it's a good idea to taste and evaluate the product first, then examine data collected from the batch record.

With sufficient experience, an operator evaluating the quality of natto is more reliable, and much cheaper, than any laboratory analytical tests. If this is conducted for every batch of natto, any abnormalities can be quickly identified, and the process can be troubleshooted if problems with the product appear.

Standard Natto Method (Descriptive Tests)

The Japanese natto industry has a standard method for the qualitative evaluation of natto, which was first described in the textbook 'Methods of Natto Research'. This method, which can be classified as a descriptive test in scientific sensory evaluation, is as follows.[55]

1. Leave the package of natto at room temperature for 1h after removing from the refrigerator.
2. Remove the polyethylene film and inspect the top surface.

3. Flip the container over, and let the natto drop out onto the lid, and inspect the bottom surface.
4. For steps 2 and 3, inspect the surface colour, bacterial coverage, etc.
5. To evaluate stickiness, stir the natto 20 times with disposable chopsticks, then lift one-third to one-half of the natto about 40cm.
6. Smell the aroma, and taste the beans, evaluating also for hardness.

After conducting this evaluation, test subjects are instructed to complete a 5 point questionnaire. A total of 9 parameters, each with a prompt, are scored on a scale of 1–5, with 1 being 'bad' and 5 being 'good'.

Table III.4 Natto sensory questionnaire

	Evaluation Scale				
Evaluation items	1	2	3	4	5
1. Coverage of bacteria	Bad	–	Normal	–	Good
2. Bacteriolytic condition	High	–	Normal	–	Low
3. Cracked or crushed beans	High	–	Normal	–	Low
4. Colour of beans	Bad	–	Normal	–	Good
5. Smell	Bad	–	Normal	–	Good
6. Hardness	Hard	–	Normal	–	Soft
7. Taste	Bad	–	Normal	–	Good
8. Stickiness	Weak	–	Normal	–	Strong
9. General	Bad	–	Normal	–	Good

Evaluation Prompts:

1. Coverage of bacteria

(Good) Natto is covered with a certain thickness, no unevenness, and no lumps.

(Bad) Speckled, mottled or thin coverage of beans in some places.

2. Bacteriolytic condition

(Low) No dissolved biofilm found on the bean's surface.

(High) Natto has a sticky condition due to breakdown of the bacterial biofilm (see Note 1).

3. Cracked or crushed beans

(Low) Little to no cracked, hardened, or peeling beans.

(High) Many of the above.

4. Colour of beans

(Good) Brown hulls or light brown hulls with a bright colour.

(Bad) Dark brown or black hulls.

5. Smell

(Good) Good, sweet smell that is moderate in terms of ammonia smell, roasted smell (see Note 2), acid smell, and other odors.

(Bad) Inadequate in terms of sweetness, ammonia, roasted smell, acid, and other odors.

6. Hardness
(Soft) Soft and smooth to chew.
(Hard) Hard, and difficult to chew.
7. Taste
(Good) Appropriate taste in terms of umami, bitterness, sweetness, and other strange tastes.
(Bad) Unsuitable taste in terms of umami, bitterness, sweetness, and other strange tastes.
8. Stickiness
(Strong) The product has a high viscosity and good stringiness when it is stirred.
(Weak) The product has little stickiness and forms weak threads.
9. General
An overall score.

Some prompts may have been translated improperly.
Note 1: The original parameter is (溶菌状態), translated to bacteriolysis, and is expected to be a condition that occurs if the product temperature exceeds 55°C for extended periods of time during fermentation.
Note 2: The original descriptor is (コゲ臭), translated to 'roasted smell', but may refer to toasted, burning, etc. The evaluation criteria for "5. Smell" is subjective.

Test subjects conducting this sensory evaluation should also note down basic info, such as the product being tested, their demographic info, and whether they consume natto regularly or have received formal training on the matter.

Different testers will have different preferences, so testers for new products and research settings have a large panel of evaluators. It is also best practice to provide a 'standard' package of natto from the supermarket as a baseline and assign this natto a score of 3 throughout. When evaluating a large number of samples, the evaluator should take measures to avoid taste fatigue. Dilute *shoyu* is cited as being an acceptable method of cleansing the palate during a large number of tests.[55]

Difference Tests & Affective Tests

In addition to the natto sensory questionnaire, natto producers and researchers will also employ other common sensory evaluation techniques that are described in food science. One of the most widely used types of tests are called **difference tests**, in which each test subject will evaluate two or more samples, and see if they can tell the difference. Another category of tests are **affective tests**, in which the test subject will be asked to evaluate two or more samples and pick whichever one they prefer.

For example, let's say a natto company breeds a new strain of *Bacillus subtilis* and wants to evaluate the resultant natto. They can conduct a type of test called a **triangle test**, a form of difference test, in which the test subject will be given three samples: one of which is made with the new strain of *B. subtilis*, and the other two are regular commercial natto. Then, they will be asked if they can identify the different one.

The researcher might also conduct a **paired preference test**, which is a form of affective test, in which they will be given two natto samples: one made with the new strain, and one regular commercial natto, and asked which they prefer. This can be made more specific if using the **attribute-by-preference test**, in which the test subject is asked which sample is preferrable in terms of a certain attribute, like stickiness or aroma.

Different tests will be used depending on the objective of the evaluation. A summary of the different types of tests is provided below, and a full list of test types, and the details of how they are conducted, is provided in literature.[56,57] These tests are used extensively in research and development for new product lines in the food industry.

Table III.5 Types and examples of sensory tests

Test Type	Objective	Example
Difference Test	Does a difference exist between samples?	Triangle test: identify the natto sample that is different among three
Attribute Difference Test	How does a certain attribute differ between samples?	Attribute-by-preference: identify the natto sample with the better aroma
Affective Test	Which sample is preferrable or acceptable?	Paired preference: pick the sample of natto you prefer
Descriptive Test	Rate each attribute on a scoring sheet	Complete the natto sensory questionnaire

3.2 Natto-Ren and National Contests

Natto Kanpyokai

In true Japanese fashion, there is an annual national contest called *Natto Kanpyokai* (lit. Natto Contest) organized by the Japan Natto Cooperative Society Federation (Natto-Ren). This event, spectacularly filled with prefectural mascots and judges in lab coats, features things like (from the 2020 *Natto Kanpyokai*):[58]

1. *Nobi-nobi* championship [のびのび選手権]–the maximum length a package of stirred natto can be stretched (in 2020, the winner was 16.34m)
2. *Maze-maze* championship [まぜまぜ選手権]–the maximum number of times you can stir a bowl of natto in one minute (in 2020, it was 136 times)
3. Creation of an ultra-long *natto-maki*

The centrepiece of this event is the natto appraisal, in which natto producers from across Japan submit their best products, which are evaluated by a panel of judges on three factors: appearance, aroma, and taste/texture, each scored out of 5. In 2020, there were 189 entries from 73 manufacturers, evaluated by 31 judges. There is a single grand prize, and separate divisions for different grain sizes.

The Natto Shinuchi Exam

If you reside in Japan, you can also take an annual exam called the *Natto Shinuchi*, which is scored out of 100 points and covers all aspects of natto history, production, and science. Based on your score, you can be given a rank based on the traditional *Rakugo* promotion system.[59]

1. 90 points or more: *Natto Shinuchi* (you get a commemorative badge)
2. 75 points or more: *Futsume* (you get a keychain)
3. 55 points or more: *Zenza* (sense of pride and accomplishment)

The *Shinuchi* with the highest scores get to participate as judges in the *Natto Kanpyokai* that year.

3.3 Quantitative Evaluation: Lab Methods

For most of the parameters in the qualitative natto questionnaire, there is a corresponding quantitative measurement that can be taken with analytical equipment. This equipment tends to be expensive, so such tests will only ever be performed in large scale natto manufacturing facilities or in the context of research, development, and in academia. These analytical methods are described in detail in 'Methods of Natto Research',[55] and 'Investigative Approach to Natto',[37] as well as standard food analysis texts.[60]

Analytical methods provide quantifiable characteristics of a natto sample, as well as the properties of natto during different stages of the production process. Aside from quality evaluation, they are also used for nutrition labelling, food safety (testing for hazards), and validating suppliers. In academic studies of the natto-making process, detailed analyses are also performed at every major step of the process to track the changes in soybean's composition and properties over time.

Some of these methods are standardized within the food industry, so I'll highlight some of the ones unique to natto, as well as those that are typically employed in normal natto production.

Table III.6 Typical analytical methods for natto at different stages.

Testing Parameter	Testing Method & Notes	Ref.
Raw soybeans		
Dimensions*	Sieve size, measurement along three axis-can be used to calculate volume and shape index	37,55
Seed coat colour	Colorimeter: L*a*b* (CIELAB colour space) typically used	34
Cotyledon colour		
% Intact grains	No. of healthy grains out of 100	34
% Damaged grains	No. of damaged grains out of 100	34
% Hardened grains	No. of grains that remain hard after soaking, out of 100	34
100-grain weight	Weight of 100 grains	34
Void ratio	Fluid paraffin method, bulk density calculation	34
Moisture content*	Moisture analyzer (105°C drying)	60
Germination rate	No. of grains that germinate out of 100-may not affect natto quality but is an indication of storage conditions. Moisten beans, germinate at 25°C, 3 days	34
Soaking		
Seed coat lift rate	No. of grains in which seed coat separates from cotyledons when soaked, out of 100; soaking 25°C, 15mins. High lift rate may adversely affect product quality and filling	34
Water absorption ratio*	Can be calculated by: Ratio = W_1 (after soaking) / W_0 (raw soybeans) For higher reproducibility, multiply W_0 by raw soybean dry matter (1–moisture content).	34
TDS in soaking water	Filter, determine solids in filtrate	34
pH of soaking water*	Low pH indicates proliferation of lactobacillus	60

Testing Parameter	Testing Method & Notes	Ref.
Steaming		
Hardness of beans*	Crushing force. Natto producers use custom apparatus (see Ref.)	[37,55]
Steamed bean weight ratio*	Can be calculated by: Ratio = W_2 (after steaming) / W_1 (after soaking) For higher reproducibility, multiply W_1 by raw soybean dry matter W_0(1–moisture content). Can be measured during production process by allocating a known weight of soaked beans to a stainless steel colander inside the steamer.	[34]
Colour	Colorimeter: L*a*b* (CIELAB colour space) typically used	[34]
Finished Natto		
Hardness*	Crushing force; same as previous. Natto producers use custom apparatus (see Ref.)	[37,55]
Colour	Colorimeter: L*a*b* (CIELAB colour space) typically used	[34]
Viscosity (Stickiness)	Rapid Visco Analyzer; typically natto is too viscous to measure as-is and may require adding 1g water to 10g natto to reduce nascent viscosity	[34]
pH*	Requires adding an equivalent mass of deionized water. Method 1) mix and measure pH; Method 2) homogenize at 15000RPM, 3mins, and measure pH	[45]
Moisture content*	Moisture analyzer (105°C drying)	[60]
Ammonia-N*	Various methods	[60]
Biofilm components	Poly-γ-glutamic acid, levan concentrations	[61]
Smell	Gas chromatography (GC), GC-mass spectrometry (GCMS), GC-Olfactory	[60]
Composition		
Moisture content*	Moisture analyzer (105°C drying)	[45,60]
Ash Content	Incineration (500°C–600°C muffle furnace)	[45,60]
Macronutrients	Carbohydrates, proteins, fats, breakdown of each; Various methods, AOAC methods	[45,60]
Minerals	Various methods	[45,60]
Vitamins	Various methods, Vitamin K2 is notable	[45,60]
Biological Analysis		
Bacteria count	Various methods; serial dilution and plating, counting under microscope.	[37,55]
Sporulated count	Various methods; boiling, serial dilution and plating, counting under a microscope	[37,55]
Bacteriophage count	Various methods	[37,55]

Notes:
*Parameters commonly measured in normal natto production
Measurements can be completed at any stage of production depending on objective

These methods are not exhaustive, and only a handful of measurements are taken during normal natto production, along with the qualitative natto questionnaire. The other methods are only measured in detail for research and development, calculating nutritional characteristics, or verifying supplier specifications. Many of the analytical methods listed in this table have acceptable alternates. In research settings, methods used by other researchers are often cited, and some methods change as analytical methods and equipment improve. These methods are used extensively in the various studies and experiments described in Chapters IV, V, and VI.

4. Market and Industry Trends of Natto in Japan

The Japan Natto Cooperative Society Federation keeps an accurate record of all aspects of the natto market. These include the state of natto factories, per-household spending on natto, consumption by prefecture, costs, and preferences by age group, among many others.

Natto Factories

The total number of natto factories has been declining since 1999. This is somewhat expected: as described in previous chapters, larger factories benefit from economy of scale, and with a highly optimized and uniform method for making natto, the industry favours a smaller number of larger factories.[48]

Table III.7 Number of natto factories in Japan, 1999–2018

Year	Number of Factories	Year	Number of Factories
1999	720	2009	641
2000	723	2010	591
2001	692	2011	600
2002	682	2012	580
2003	677	2013	552
2004	677	2014	530
2005	670	2015	508
2006	661	2016	489
2007	662	2017	475
2008	642	2018	458

Japanese Natto Market Size

Despite the smaller number of total factories, the size of the natto market has increased during the same period, suggesting that factories have been able to become more efficient and increase total throughput.[48]

Table III.8 Total natto market size in Japan, 2010–2019

Year	Total Market (Billion Yen)
2010	180.0
2011	173.0
2012	179.9
2013	196.1
2014	197.7
2015	202.8
2016	218.4
2017	231.3
2018	249.7
2019	250.3

Natto Consumption

Per-household (two people or more) natto consumption remains high. In the last two decades, households spent around 4000 yen per year on natto, equivalent to 120 individual 50g packs.[48]

Table III.9 Annual household (>2 persons) expenditure on natto, Japan, 2000–2019

Year	Yen per household	Year	Yen per household
2000	3740	2010	3312
2001	3902	2011	3295
2002	4200	2012	3333
2003	4068	2013	3479
2004	4119	2014	3417
2005	3914	2015	3640
2006	3891	2016	3835
2007	3867	2017	3949
2008	3744	2018	4232
2009	3469	2019	4238

Price of Natto

The price of natto has held steady, and even decreased somewhat between 2002–2018. Without adjusting for inflation (which is notably low in Japan throughout this time period), retail prices for natto surveyed in the special wards of Tokyo hovers around the 95yen mark for a 3x45g to 3x50g pack.[48]

Table III.10 Average retail cost per pack (3x50g and 3x45g) natto, Tokyo, 1999–2018

Year	Yen per pack	Year	Yen per pack
1999	100	2009	118
2000	98	2010	112
2001	104	2011	111
2002	144	2012	101
2003	137	2013	97
2004	136	2014	97
2005	134	2015	96
2006	133	2016	95
2007	133	2017	96
2008	125	2018	95

Frequency of Natto Consumption

Natto remains broadly popular across all age groups, with around 40% of those surveyed consuming natto every 2–3 days, or more frequently.[49]

Table III.11 Natto consumption frequency, surveyed 2021

		Age Group				
Frequency	2021 Average	20s	30s	40s	50s	60s+
Daily	16.8	13.5	16.6	15.6	15.4	21.4
Every 2-3 days	23.3	18.2	21.1	23.4	27.2	24.8
Every 4-5 days	9.9	9.2	11.1	10.6	7.2	10.7
Every week	13.2	12.5	12.6	15.0	11.6	13.7
Every 2 weeks	7.3	9.6	10.1	7.6	5.4	4.6
Every 3 weeks	5.2	9.2	4.6	4.3	4.6	4.0
Less than above	12.8	14.9	13.1	11.5	13.9	11.5
Never	11.8	12.9	10.8	11.9	14.7	9.2

Preferences by Soybean Size

Consumers prefer small grain natto, followed by medium grain, extra-small grain, *hikiwari*, and lastly large grain. This is the same across all age groups, although there are slight differences in different geographic regions.[49]

Table III.12 Natto preference by grain size, 2007–2021

Year	X.Small	Small	Med*	Large	Hikiwari	No pref.
2007	13.1	49.7	N/A	15.2	7.4	14.5
2009	11.3	54.1	N/A	15.1	5.1	14.4
2011	12.5	38.9	24.1	6.9	5.3	12.3
2013	11.9	36.5	24.2	8.3	7.2	12
2015	8.7	36.7	25.9	7.5	7.8	13.3
2017	13.1	36.8	23	7.9	7.6	11.6
2019	11.9	36.4	24.8	6.7	8.3	11.8
2021	12.6	38.3	21.2	7.2	9.7	11.1

*Medium category was defined in the 2011 survey, and previously overlapped with Small and Large categories. Extra-large category not defined in this survey.

Other Developments in the Natto Industry

Natto producers are all too aware that there is a subset of the Japanese population, and large portion of the global population, that does not enjoy natto, especially due to its texture and smell. At the same time, increasing health and environmental consciousness among consumers has led to a change in people's habits. Both factors have led to the development of new products in the natto industry.

For the soybean, this means the use of more diverse varieties of soybeans, including small black soybeans, which were commercially introduced for natto in 2004 following a multi-year cultivation project starting from just 50 stored seeds.[62] Black soybeans are appealing to consumers due to their anthocyanin content, but small black soybeans had not been grown in Japan since the 1970s.

For the natto bacterium, producers have developed low odour natto (introduced in 2002) and low stickiness natto (introduced in 2017). These involve breeding specialized strains of *B. subtilis*, which will be discussed in detail in Chapter IV.

Summary

The natto industry remains strong in Japan. Like many other food industries, the natto market has been consolidating into larger factories, while costs for commercial natto remains low.

There is a huge diversity in soybean types, packaging, sauces, condiments, accompaniments, and cooking methods to suit individual consumer's preferences. New products continue to be developed to meet consumer trends, demands, and to accommodate those that are less likely to purchase natto.

Qualitative evaluations and quantitative measurements can be performed on natto, for which there are several standard methods for each. Some of these tests are performed during regular natto production as a quality assurance step, while others are performed in research and development settings when testing new natto products.

The development of new natto products is essential to reaching new markets and pushing innovation in the somewhat stagnant natto industry. However, to understand these developments, we first need to understand the science of the natto making process. These will be examined in the following Chapters IV and V.

Further Reading

Sensory evaluation and food analysis are fascinating subjects, and a lot of interesting science and statistics goes into these topics. For more details, consider the below, or any equivalent thereof:

- **Sensory Evaluation Techniques** by Morten C. Meilgaard, Gail Vance Civille, and B. Thomas Carr[56]

I encourage the reader to poke around Natto-Ren's website, the organization responsible for collecting market data, organizing events, and promoting the Japanese natto industry:

- **www.natto.or.jp**

It's hard to describe the full diversity of natto products in Japan. One fantastic online resource just to browse a seemingly endless list of natto products is:

- @Taiji141's natto wiki, as mentioned previously, (**seesaawiki.jp/w/taiji141**) as well as his Instagram @Taiji141

IV. Natto Bacteria

Shin Sawamura and Shunsuke Muramatsu first isolated and characterized the bacterium responsible for natto in 1905–1906, building on the work by Kikuji Yabe a decade prior (see Chapter I).[12] Sawamura named the organism *Bacillus natto*, although later methods of classification and taxonomy would put the bacterium under the species *Bacillus subtilis.* Researchers conducting further work on the natto-producing bacterium would refer to it as *Bacillus subtilis* var. *natto* to distinguish it from other strains.

B. subtilis is a model organism and one of the most widely studied gram-positive bacteria, analogous to *E. coli*, which is the most studied gram-negative bacterium. For researchers working with natto, this means that there is a wealth of knowledge on the inner workings of *B. subtilis*, ranging from its ecology and cell biology, down to the inner workings of its molecular biology and genetics.

The applied knowledge gained from studying *B. subtilis* include methods for making spore suspensions of *natto-kin*, breeding new strains of *B. subtilis* with unique natto fermentation characteristics, studying its nutritional and physiological characteristics, and studying its relationship to food safety.

Understanding the biology of *Bacillus subtilis* is not necessary for the average natto producer, and facility operators shouldn't be expected to know the details presented in this chapter. However, these details come in handy when communicating facts about natto to both regulators and consumers, especially regarding safety and nutritional claims. It also explains some fundamental questions, such as why natto contains the highest concentration of Vitamin K_2 in known food sources, or why natto contains a fibrinolytic enzyme called nattokinase that can dissolve blood clots. As a bonus, the biology of *B. subtilis* is fascinating.

In this chapter, we'll delve into the details of one of two organisms behind natto, *Bacillus subtilis.* We'll look at its ecology–where it's found in nature–and how it was first discovered during the golden age of microbiology. We'll also look at the cellular biology of *B. subtilis*, describing the various functions of the bacterial cell, and how they behave in multicellular communities. Finally, we'll look at the genetics of *B. subtilis*, how it can be bred and modified to yield different strains of natto-producing *B. subtilis*, with some examples of strains that have been developed by researchers.

Institutions are aware that science is becoming inaccessible to non-specialists as scientific fields get more complex with modern tools and research methods. As early as 1992, the late Donald P. Hayes of Cornell University commented that academic articles were becoming increasingly difficult to understand, with topics in biochemistry and biology carrying the highest level of lexical difficulty.[63] To reach a compromise, this chapter will skip over some of the highly technical details and concepts surrounding *B. subtilis*, but still be approachable to a reader with a decent understanding of biology and biochemistry.

For those seeking more details on each topic, or for those requiring a more basic understanding of the underlying concepts, see the 'Further Reading' section at the end of this chapter.

Finally, keep in mind that the field of science is constantly evolving. As of writing this chapter in 2022, there are still gaps in the knowledge of natto-making and *Bacillus subtilis*, and some concepts are constantly changing or being revised.

1. Ecology of *Bacillus subtilis*

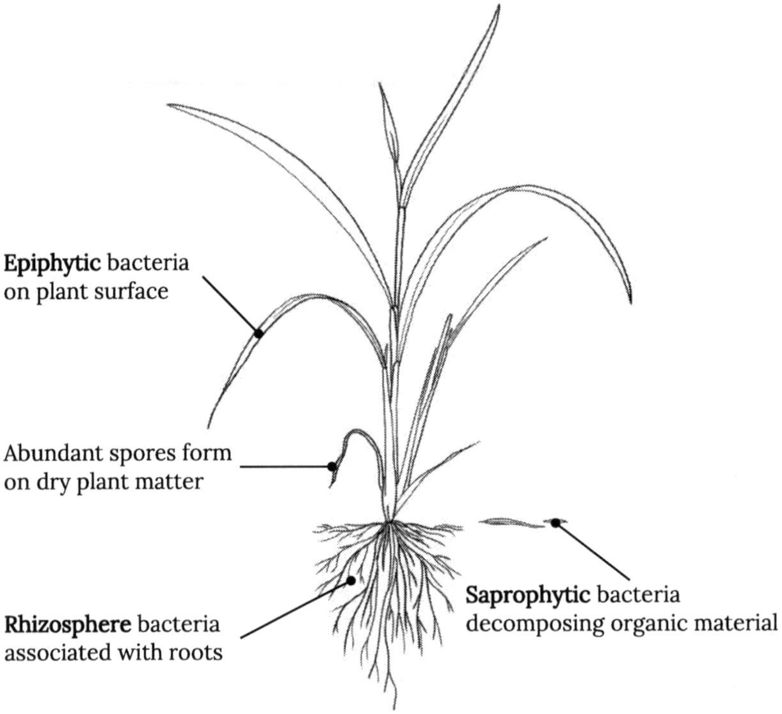

Figure IV.1 Terrestrial ecology of Bacillus subtilis

Bacillus subtilis is commonly referred to as 'hay bacillus', since early researchers working with the microorganism could easily isolate *B. subtilis* from boiled hay infusions. This is possible because *B. subtilis*, when subjected to unfavourable environmental conditions such as those found on drying plant matter, will readily form an **endospore**. This endospore contains all the genetic information and material needed to germinate and grow as a normal (vegetative) cell when conditions are favourable again. These endospores are extremely hardy, and can survive heat (including boiling water at atmospheric pressures), high levels of radiation, oxidizing agents and enzymes. Spores have been demonstrated to survive for hundreds of years, though they may last even longer.[64]

Endospores are lightweight and easily dispersed by the wind, and may become airborne and land in a hostile environment without ever germinating. Thus, it can be difficult to distinguish between finding *B. subtilis* in a certain environment, and demonstrating that it grows in this environment.[65]

Our current understanding of *B. subtilis* ecology is that it grows in **aerobic** (oxygen-rich) soil environments and marine ecosystems. Although *B. subtilis* was historically considered an aerobic bacteria, it also has the limited ability to grow in **anoxic** environments in the presence of nitrates (NO_3^-) and nitrites (NO_2^-), which often occurs in water-logged soils, as well as **anaerobic** environments.[66]

Within soil ecosystems, *B. subtilis* is found in a **saprophytic** role, feeding on dead or decaying organic matter within soil. It's also found within the **rhizosphere**, or in close association with plant roots, where it's been demonstrated to promote plant growth through activation of host plant defence systems, outcompeting potential pathogens, and making nutrients more readily available. *B. subtilis* also grows in **epiphytic** environments on the surfaces of plants, and is commonly found on live and decaying plant leaves.[65] Lastly, *B. subtilis* has also been found abundantly in marine environments.[67]

These factors explain why natto was historically created out of soybeans wrapped in rice straw. *B. subtilis* cells residing on the surfaces of rice plants would form hardy endospores as the rice plants decay and dry during harvest. Once hot, steamed soybeans were introduced to the rice straw, the *B. subtilis* spores germinate, and their large number gives them a head start over other organisms and would eventually out-compete them. However, dried rice straw is far from being a pure *B. subtilis* culture, and natto made with rice straw could contain multiple different organisms, sometimes with no adverse effects on the product quality, but occasionally leading to failed natto.

2. History and Taxonomy

Early Discoveries

In Chapter I, we left off with Jun Hanzawa and Jiro Miura pioneering the modern natto production process using the microbiology work of previous researchers. These included Shin Sawamura, who first identified strains of *Bacillus* responsible for natto, and Shunsuke Muramatsu, who demonstrated that the qualities of natto could be brought about with a single strain of *Bacillus,* which Sawamura named *Bacillus natto.*

Neither Sawamura nor Muramatsu was the first to isolate the bacterium that would eventually be called *Bacillus subtilis.* This credit goes to Ferdinand Cohn, who described the bacterium in 1872, classifying it to the genus *Bacillus* (meaning rod-shaped, although this term not particularly useful for distinguishing between many rod-shaped bacteria) and giving it the species name *subtilis* (meaning slender).[68] Cohn's bacterium was likely identical to another bacterium isolated in 1832 by Christian Gottfried Ehrenberg, who named the bacterium *Vibrio subtilis.* However, bacteriological techniques were so primitive at the time that no one really knows for sure if the bacterium isolated by Ehrenberg was indeed *B. subtilis*. The only surviving record is a brief description of the bacterium and illustrations showing that it was a motile rod that formed long chains.

Cohn originally isolated *B. subtilis* from sour milk, and infusions of distilled water with lupines and peas. In later research, Cohn would use infusions of boiled hay for studying spore formation.

When Sawamura isolated the bacterium responsible for natto, he named it *Bacillus natto,* thinking that it was a novel strain of *Bacillus* responsible for the transformation of soybeans into natto. *Bacillus natto* would later be reclassified as a variant of the already discovered *Bacillus subtilis*, but Sawamura was not misguided in giving it the name *Bacillus natto.* During this time bacterial taxonomy wasn't well developed, and the species *Bacillus subtilis* exhibits a wide range of possible traits. Taxonomy and the definition of a species is also an entirely human-made concept, and lines between species get blurry when classifying bacteria.

Bacillus subtilis or Bacillus natto?

Nevertheless, by 1946, as part of the United States Department of Agriculture (USDA) miscellaneous publication No. 559, bacteriologists were already classifying Sawamura's *Bacillus natto* to be a strain of *Bacillus subtilis*.[69] This was further agreed upon in 1949 in the 6th edition of *Bergey's Manual of Determinative Bacteriology*, considered at the time to be the authoritative text on bacterial taxonomy. The author of the chapter on *Bacillus subtilis* aptly stated:[70]

> "The identity of this species has been the subject of some controversy owing to the indefiniteness of the original descriptions, to the distribution of cultures under the name Bacillus subtilis that were incorrectly identified, to variations in the forms of growth that may be observed, and to confusion with *Bacillus cereus.*" *(Bergey's Manual of Determinative Bacteriology, 6th Edition, p. 708–709)*

Bacillus subtilis is so ubiquitous, and can be found in so many forms, that early researchers ended up coming up with many names for what was taxonomically the same bacterium. The manual would go on to list a long list of previously identified and named strains of *Bacillus* that were 'probably identical with or variants of *Bacillus subtilis*', which included Sawamura's *Bacillus natto*, among others.

The reclassification of *Bacillus natto* into *Bacillus subtilis* irked some Japanese researchers, especially since *B. natto* had characteristics that were clearly unique to natto production, and not all strains of *B. subtilis* will create a satisfactory natto product. Specifically, *B. natto* produced **mucoid** colonies responsible for the sticky texture of natto, while some strains of *B. subtilis* could not.

To argue for one name or the other is futile. Taxonomy is a human invention used to subdivide organisms into classifications based on the needs of the taxonomer. For bacteria, this is especially difficult. All the traditional 'species concepts', which were developed for higher organisms, don't quite work for distinguishing between bacteria. For example, it's easy to classify dogs and wolves into a common genus, *Canis spp.* based on their visual and anatomical features. The same cannot be said for two microscopic rod-shaped specs, which may not even be in the same kingdom of life.

Before the development of more advanced genetic techniques, a lot of bacterial classification methods relied on subtle visual characteristics under a microscope, the morphology and colour of a colony under specified media conditions, et cetera. These would inevitably lead to taxonomical revisions once bacterial genomes were sequenced, giving researchers a better understanding of bacterial phylogeny.

The name *Bacillus natto* would stick around in Japanese research papers well into the latter half of the 20th century. By 1975, *Bergey's Manual of Determinative Bacteriology*, 8th Edition, formally considered *B. natto* to be a synonym for *Bacillus subtilis,* and would be reclassified into the latter.[71] *Bergey's Manual of Determinative Bacteriology*, now in its 9th Edition, and *Bergey's Manual of Systematics of Archaea and Bacteria*, continue to be the authoritative references for classifying bacteria today. Their methods are still useful for classifying bacterial cultures and can be done with basic laboratory equipment and materials, some of which are presented in Section 6.2.

The ubiquity of *Bacillus subtilis*, combined with the fact that it is safe and easy to study, means that it is considered the **type species** for the genus *Bacillus.* This means that *B. subtilis* is considered the 'quintessential' *Bacillus,* and represents the flagship species of the genus. Further classifications and taxonomy would revolve around *Bacillus subtilis,* and the species name would remain the same to present day. For the same reasons, *B. subtilis* is considered a **model organism** for gram-positive bacteria (in general, the phylum *Bacillota*, formerly *Firmicutes*). Our understanding of the basic mechanisms of bacteria is built off extensive studies of the gram-positive *B. subtilis* and gram-negative *E. coli,* among a handful of other model organisms.

DNA and the Phylogenetic Species Concept

The discovery of the structure and function of deoxyribonucleic acid (DNA) through work of Rosalind Franklin, Jim Watson, and Francis Crick in 1957 lead to the development of more accurate definitions of bacterial species based on their similarities in DNA. The first approaches were simple, and included analyzing guanine and cytosine (**G+C**) **content** (bacterial whose molecular percentage of G+C were clearly not the same) and using **DNA–DNA hybridization** techniques, where a double strand of DNA between two organisms would be split into single strands, and scientists would see how well complementary strains of DNA 'stick' to one another.

In the 1970s and 1980s, as DNA sequencing technology improved, researchers demonstrated that sequencing certain DNA fragments is incredibly useful for phylogenetic analysis, leading to the widespread use of the **phylogenetic species concept**.

The basis of this concept is that all organisms within each kingdom of life contain certain genes in their DNA that essential, highly conserved, and evolve relatively slowly. If we look at the variations within that gene across species, we can infer how closely related two species are on an evolutionary timescale, with the assumption that random mutations will occur periodically from a common ancestor.

Taxonomers of bacteria use the **16S ribosomal RNA (16S rRNA)** sequence for studying bacterial species, although there are some known issues with this approach.[72] 16S rRNA is a component of the 30S subunit of prokaryotic ribosomes. The 30S subunit is, in turn, the smaller portion (the small subunit, or SSU) of the bacterial ribosome. Every single organism has ribosomes, which facilitate the translation of messenger RNA (mRNA) into proteins, so the genes associated with the ribosome are good candidates for phylogenetic analysis.

By sequencing 16S rRNA, we can examine the relationship between *Bacillus subtilis*, its subspecies, and other species in the genus *Bacillus*. These are usually visualized in a **phylogenetic tree**.

Briefly, these diagrams show the similarity between species by comparing the number of substitutions in a genetic sequence (in this case, 16S rRNA) between these species. The total length of all horizontal lines connecting two species on a phylogenetic tree is proportional to the number of substitutions in that gene, which are depicted with a scale in units of substitutions per site.

Similar species are organized into clades, which are groupings on the tree showing the evolutionary relationship between species, assuming that they evolved from a common ancestor. At each node, bootstrap values (0–100) represent our confidence that the clade of species within that node are organized as such, since there may be multiple ways that clades can be mathematically organized.

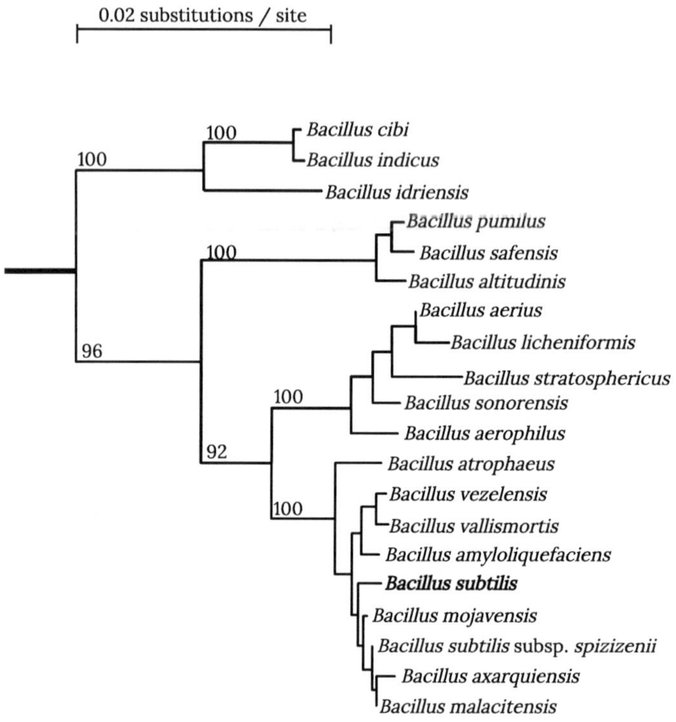

Figure IV.2 Portion of a phylogenetic tree of Bacillus spp. adapted directly from Bergey's Manual of Systematics of Archaea and Bacteria. Bootstrap values above 85 are shown.

Phylogenetic trees can give us useful insights into the definition and organization of species. In this example, we can see that *Bacillus subtilis* is closely related to *Bacillus amyloliquefaciens*, but more distantly related to the deadly pathogen *Bacillus anthracis* (not shown in the figure above).

At what point do two strains of bacteria become two different species (*B. subtilis* vs *B. natto*) or two different subspecies (*B. subtilis* vs *B. subtilis* var *natto*)? We'll leave this problem to bacterial taxonomers and the scientific census. To reiterate, taxonomy is a human-made concept that allows us to neatly organize 'species' according to our own needs. The bacterium itself will simply function based on its own genetics.

3. Cellular Biology of *Bacillus subtilis*

3.1 Cellular Functions and Reproduction of *Bacillus subtilis*

In *Bergey's Manual of Systematics of Archaea and Bacteria*, the bacterium *B. subtilis* is characterized as aerobic, gram-positive, motile, and rod-shaped, forming ellipsoidal to cylindrical spores. Each cell measures 0.7–0.8μm in diameter and 2.0–3.0μm in length.[73]

B. subtilis is **gram-positive**, meaning that it stains violet through the process of **gram staining**, whereby cells are subjected to a crystal violet stain and iodine solution, washed with a solvent, then observed under a microscope.

Gram-positive bacteria such as *B. subtilis* retain the crystal violet stain and show up violet after gram staining, whereas in gram-negative bacteria such as *E. coli*, the crystal violet is washed away and only the pink-coloured counterstain remains.

This method organizes bacteria into two large groups: the gram-positive and gram-negative bacteria, distinguished by the thickness of their **cell walls** and the organization of their **cell membranes** and cell walls. Gram-positive bacteria such as *B. subtilis* have a single cell membrane enveloped by a thick layer of peptidoglycan, whereas gram-negative bacteria such as *E. coli* have a thin layer of peptidoglycan sandwiched between two membranes. These two morphologies have a lot of implications for how the cells operate. For example, only single-membrane bacteria such as *B. subtilis* can form endospores, while double-membrane bacteria such as *E. coli* are more resistant to antibiotics.

Like many other concepts in biology, there are a lot of exceptions. The gram stain does not necessarily define the structure of the cell membrane and cell wall, and there are some species that are gram-variable or gram-indeterminate, even if they have one of the cell wall and cell membrane structures above. Bacteria with a single membrane and thick cell wall are classified under the phylum *Bacillota* (formerly *Firmicutes*).

Table IV.1 Scientific classification of Bacillus subtilis var. natto

Domain:	**Bacteria**	Family:	**Bacillaceae**
Phylum:	**Bacillota**	Genus:	**Bacillus**
Class:	**Bacilli**	Species:	**B. subtilis**
Order:	**Bacillales**	Subspecies:	**B. subtilis** subsp./var. **natto**

The Bacillus subtilis Cell

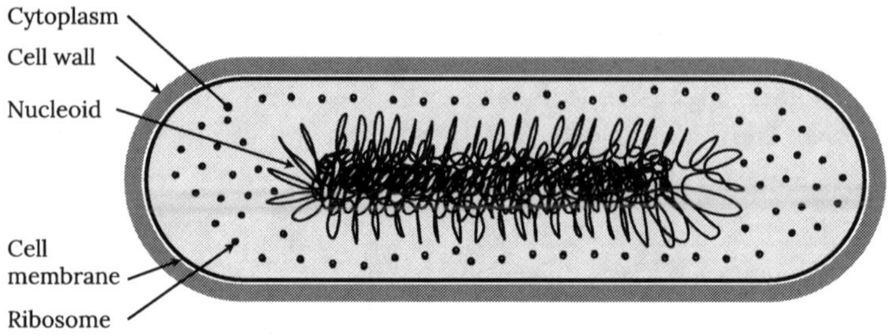

Figure IV.3 Anatomy of a Bacillus subtilis cell

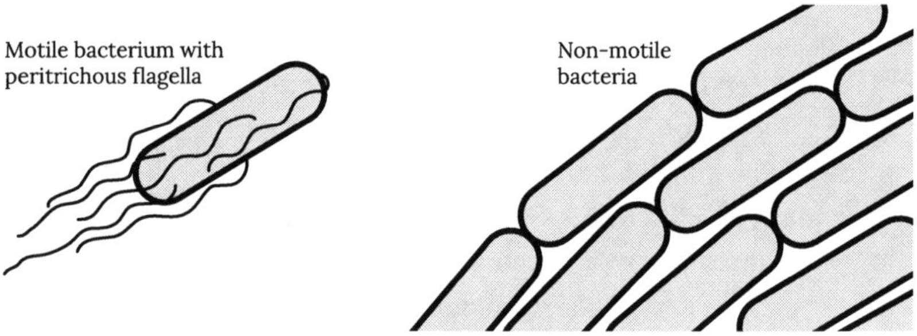

Figure IV.4 Motility in Bacillus subtilis

The **cytoplasm** of a vegetative *B. subtilis* cell is surrounded by the cell membrane (or cytoplasmic membrane), which is surrounded by a thick cell wall composed of mainly peptidoglycan (also called murein). Depending on the strain of *B. subtilis*, the entire cell wall may also be contained inside a **capsule**, although this often overlaps with the biofilm.[74]

Some strains of *B. subtilis* have **peritrichous flagella**, which are long, hair-like structures that protrude out in all directions and give the bacterium **motility**. However, not all strains of *B. subtilis* have this trait, and even ones that can form flagella will not do so in certain conditions.

Dispersed within the cytoplasm, the cytoplasmic membrane, the membrane-cell wall interface, the cell wall, and the medium itself is a collection of proteins called the **proteome,** which carries out the machinery and biochemistry of the cell.[75]

These proteins are coded by a single **chromosome** organized into a ring which measures 4.2 million base pairs in length. Fully stretched out, this ring would be around 1.5–2mm in diameter, but inside the cell it is compacted into a 1–2µm mass called the **nucleoid.**[76] A small number of proteins may also coded by **plasmids**, which are much smaller rings of DNA separate from the main chromosome.

Like all prokaryotes, *B. subtilis* lacks any organelles, and contains no nuclear envelope. **Ribosomes** can move freely within the cytoplasm, translating DNA into proteins with the help of **tRNA**. Through a variety of different mechanisms, these proteins end up inside the cytoplasm, the cytoplasmic membrane, the membrane-cell wall interface, the cell wall, or the extracellular medium, depending on their function. We'll look at the proteome and molecular biology of *B. subtilis* in detail in Section 5.

Cell Growth and Binary Fission

Under conditions favourable for growth, a single *B. subtilis* cell will elongate, growing along the centre of its body (A: see page 114). If conditions are right, the cell will replicate its chromosome, with copies located at opposite sides of the dividing cell, and the cell will split into two identical cells in a process called **binary fission.**

This process is orchestrated by an ensemble of membrane associated protein complexes, and a protein called FtsZ, which is normally dispersed within the cytosol. To summarize, FtsZ forms a structure called the Z-ring in the interior of the centre of the cell (B). This Z-ring attracts a host of other proteins to the division site, most of which are associated with the cell membrane, and together they form the **divisome** (C). The newly constructed divisome promotes the formation of a new cell wall and the pinching-off of the cell membrane (D), eventually forming a septum, and the divisome is disassembled back into its parts.[77]

The two daughter cells will tend to stick to one another, joined by the septum peptidoglycan, until murein hydrolases released from the cell have time to break the inner part of the septum (E). This 'stickiness' means *B. subtilis* cells and other bacteria often occur in pairs, and during very high growth rates, they can occur in chains.[77]

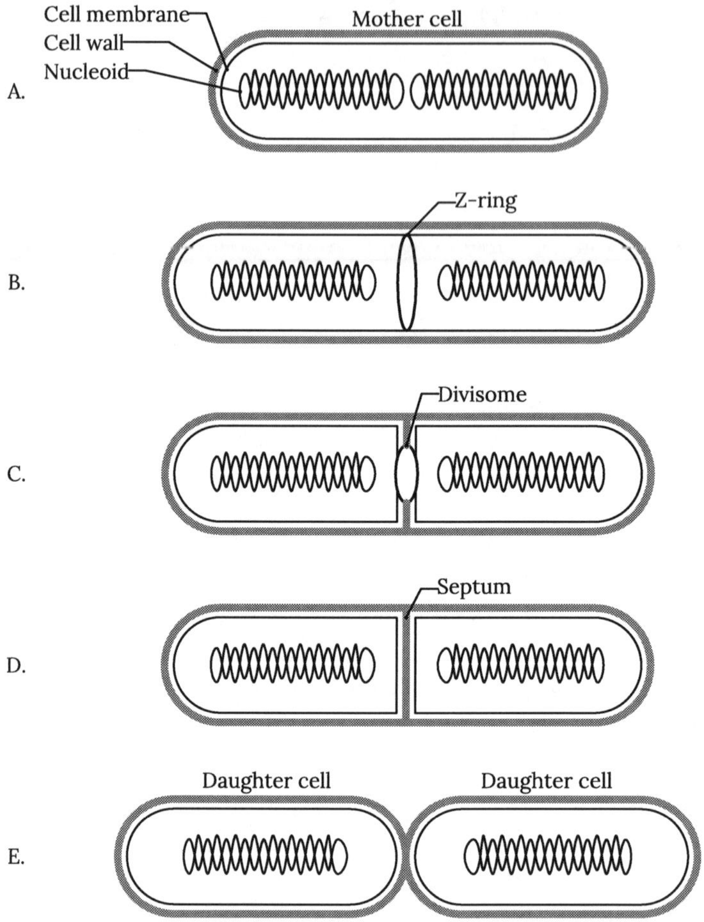

Figure IV.5 Binary fission in Bacillus subtilis

Sporulation

If conditions are adverse, such as the lack of food (carbon, nitrogen, or phosphorous sources), water, or the presence of certain compounds, *B. subtilis* will undergo sporulation, leading to the formation of a resilient, dormant endospore. The process begins like binary fission, with the formation of a Z-ring at the centre of the cell. However, this Z-ring won't form a divisome. Instead, the Z-ring morphs into a spiral like structure and migrates to opposite cell poles, eventually forming two Z-rings (Stage I).

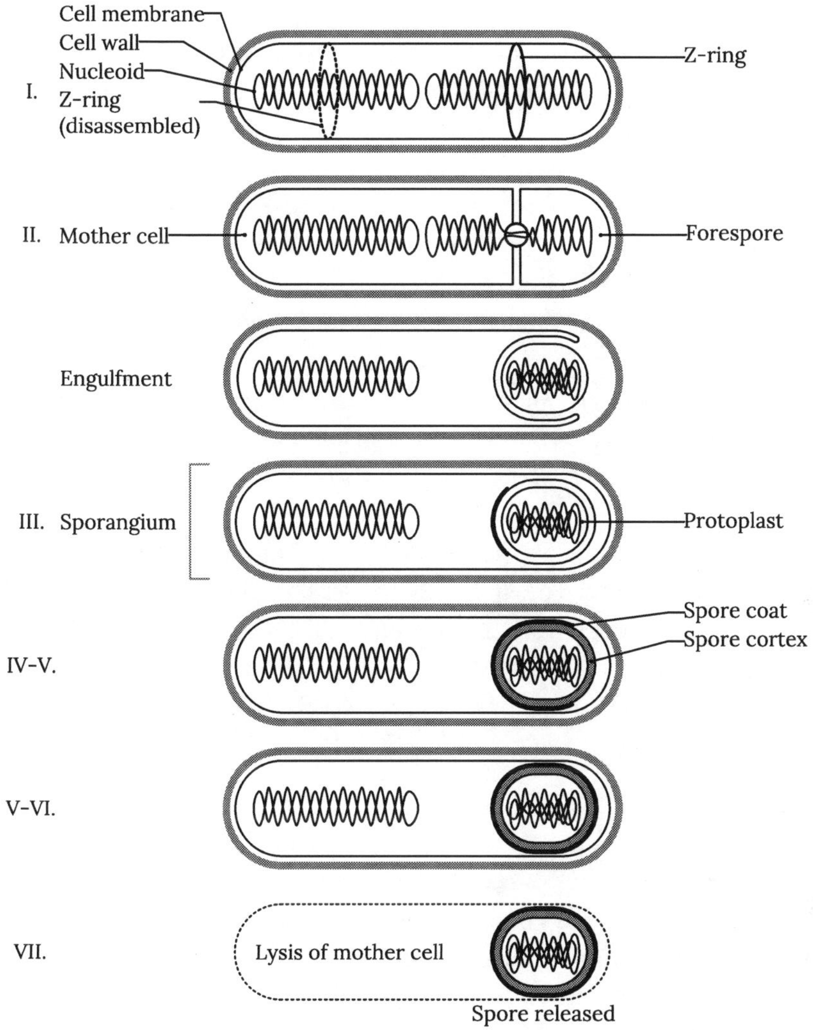

Figure IV.6 Sporulation in Bacillus subtilis

One of these Z-rings is disassembled, and the other will form the new divisome for sporulation. This divisome splits the pre-divisional cell into a larger mother cell, and a smaller daughter **forespore** (Stage II). This uneven division traps the chromosome copy that is designated for the forespore, so the cell must use a translocase enzyme to 'pull' the rest of the chromosome into the forespore.

In the next stage, the mother cell fully engulfs the forespore, creating a double-membrane protoplast within the mother cell (Stage III). Once engulfed, two protective structures are formed around the forespore: the **spore cortex** and the **spore coat**. The spore cortex is a modified form of peptidoglycan that sits atop the forespore's thin peptidoglycan cell wall, between its inner and outer membranes (formed between Stages III through V). The spore coat is a shell of over 70 different proteins that forms multiple layers of protection around the forespore (formed between stages III through VI).

Finally, the mother cell lyses and dies, releasing the spore into the environment (Stage VII). The spore coat will continue to mature shortly after this event through cross-linking of spore coat proteins, which increases its strength.[77]

When looking at the double-membrane protoplast formed after engulfment, scientists can't help but to wonder if double-membraned, gram-negative bacteria arose from sporulating single-membrane gram-positive bacteria. There is a fascinating theory that all gram-negative bacteria arose from spore-forming gram-positive bacteria, since the latter could better survive a cataclysmic extinction event.[78]

Spore Germination and Outgrowth

The spores of *B. subtilis* will remain in their dormant state until conditions are favourable again. Favourable conditions are detected by the spore through a wide range of compounds known as **germinants**, which include sugars such as D-glucose or amino acids such as L-alanine that signal to the spore: "wake up!". Once the spore detects the germinant there's a brief period where you can remove the germinant, and the spore won't germinate. However, after this period has elapsed, the spore is committed to the irreversible process of germination.

During germination, the spore is hydrated, and the cortex is hydrolyzed, leaving behind only the spore coat, and the core (protoplast). In the initial stages, the spore depends on internal energy stores for growth. Some of the spore's proteins are also broken down into amino acids, which can be used as resources for the germinating spore. At this point, the spore no longer has any of its resilient properties. Metabolism, RNA, and protein synthesis starts.

Finally, the spore coat ruptures, and the new vegetative cell emerges, in a process called **outgrowth**. The whole process takes around 60–100 minutes.[79]

3.2 Motility and Multicellularity of *Bacillus subtilis*

So far, we've examined the structure and reproduction of individual *B. subtilis* cells. If these cells are introduced to a suitable environment, they will exhibit behaviours that allow them to quickly colonize that environment and survive. These behaviours include **motility**: its ability to move, and the formation of a **biofilm**: where many cells grow together bound by extracellular substances as a strategy for survival.

So called 'wild type' strains of *B. subtilis* tend to exhibit a wider range of possible behaviours to allow them to adapt to many environments. In domesticated strains used in lab environments and commercial natto starter, some of these strategies have been affected–more on that in Section 4.

Swimming Motility and Taxis

B. subtilis can swim in a liquid environment through its peritrichous flagella. It will tend to swim towards areas that favour growth, a process known as **taxis** (e.g. chemotaxis, to swim towards areas of high concentration of a particular compound, or aerotaxis, to swim towards areas of higher oxygen).

It achieves this by using a technique called 'tumble-and-run', which was first observed in *E. coli* in the 1960s.[80] If *B. subtilis* rotates all of its flagella counterclockwise, the flagella will form a bundle and cause the bacteria to swim smoothly in one direction, or 'run'. If it rotates any of its flagella clockwise, these flagella will unwind from the bundle, causing the bacteria to 'tumble', which randomly reorients the bacterium.[81,82]

'Bias' refers to the bacterium's tendency to rotate its flagella counterclockwise and run. If the bacterium senses that it is in a favourable environment, it will increase its bias and tend to run rather than tumble.

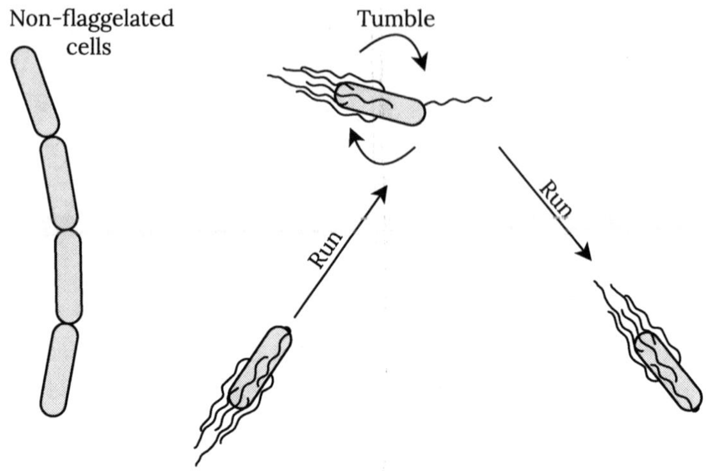

Figure IV.7 Swimming motility and taxis

Conversely, if the bacterium senses it is in an unfavourable environment, it will lower its bias and tend to tumble rather than run. The net effect of this simple mechanism is that the bacterium will exhibit a random walk that explores its environment and tends to migrate towards areas with favourable conditions.

Taxis shows up across many distantly related species of bacteria, since it's such a useful trait for survivability, and highly conserved.

Swarming Motility

B. subtilis can quickly migrate across a substrate surface using swarming motility. At the leading edge of a colony, individual cells associate side-by-side with one another in rafts that become hyperflagellated. The entire raft moves in one direction, although separate rafts are free to move independently. Swarming bacterium can colonize a petri dish of soft agar faster than individual swimming bacteria. Scientists theorize that this motility strategy reduces drag on the combined raft.

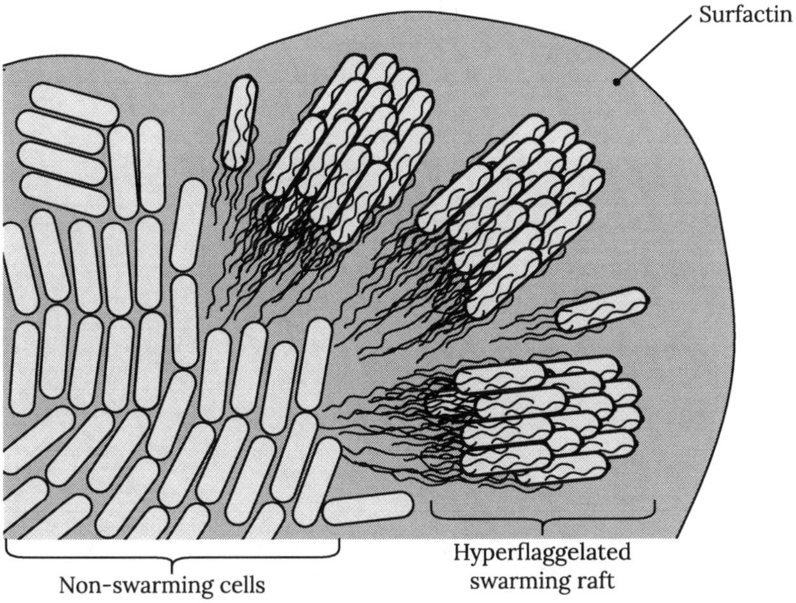

Figure IV.8 Swarming motility

Swarming motility requires **surfactin**, which is a lipopeptide secreted by the bacterium. Just like detergent, surfactin is **amphipathic**, meaning that it's a molecule with both a polar and non-polar end. This allows it to break the surface tension of water, greatly improving the colony's ability to spread out along a wet surface. Surfactin also acts as a potent antibiotic compound against other competing microorganisms.[82]

Sliding Motility

It's also possible for *B. subtilis* to migrate across a surface simply by reproducing. The growing colony pushes cells outwards with the aid of surfactin, migrating across a surface through a method called sliding motility. This may not appear to be true 'motility', but the strategy is effective enough that proper sliding motility can colonize an entire petri dish of soft agar in 16 hours,[82] which cannot be achieved by bacteria lacking these traits.

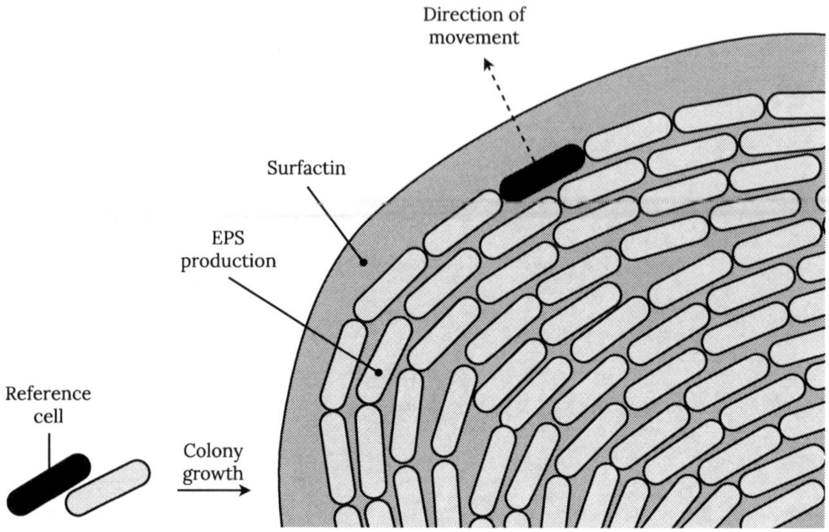

Figure IV.9 Sliding motility

Sliding motility is aided by the formation of a biofilm, which is a mixture of extracellular polysaccharides, proteins, and other substances secreted by the colony. The substances of the biofilm are hygroscopic, and absorb water from its surroundings, hydrating the colony.

Swimming and biofilm formation represent antagonistic motifs of growth in a population of *B. subtilis*. Through several mechanisms, bacteria will switch from flagellated swimming to non-motile growth in chains or biofilms.

Biofilm and Aerial Structures

Once a growing population of cells reaches a certain density, they will start to organize themselves in structure known as a biofilm. In a standing liquid culture, this biofilm forms on the surface of the liquid and is known as a **pellicle.**

The components of the biofilm vary among different strains of *B. subtilis.* The substances secreted by, and surrounding the bacteria are collectively referred to as extracellular polymer substances (**EPSs**), or exopolymeric substances. These substances enhance cell-to-cell and cell-to-surface adhesion, absorb ions and water, and protect the colony from external conditions, other microorganisms, and bacteriophages. They also act as a storage for food and nutrients under favourable conditions and can be broken down for later use. Two EPSs, levan, and poly-γ-glutamic acid, are the principal components of the biofilm in natto.

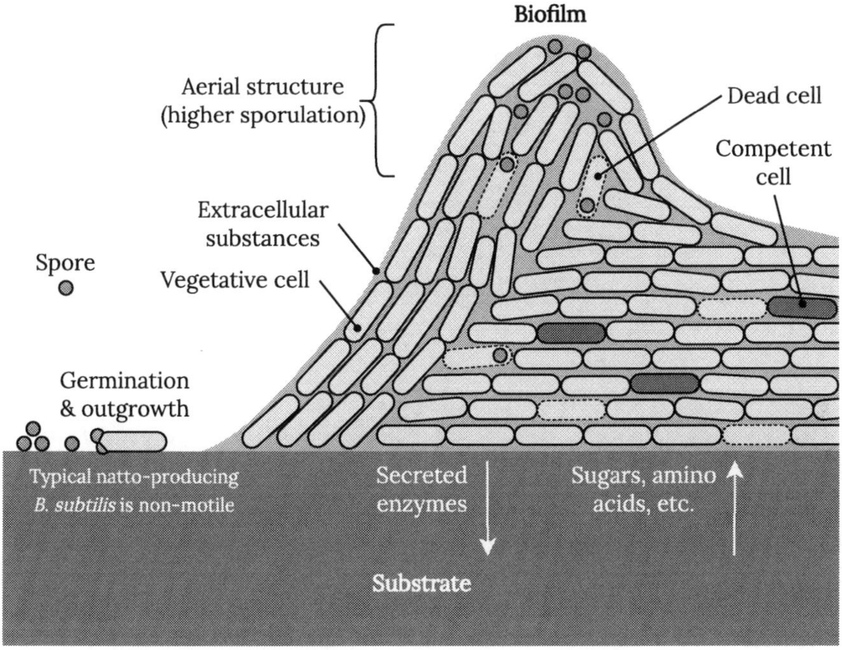

Figure IV.10 Biofilm of bacillus subtilis

As the biofilm matures it can form wrinkled structures that protrude into the air. Sporulation in a biofilm tends to take place in these aerial structures, so they are often referred to as fruiting bodies. This strategy helps with the dispersion of endospores, like the way fungi form fruiting bodies to disperse their spores, but in a more primitive fashion.

During sporulation, *B. subtilis* has also been observed to exhibit fratricidal and cannibalistic behaviour. Sporulating cells will produce toxins that kill nearby non-sporulating sibling cells. This releases additional nutrients into the biofilm, which delays the sporulation of other cells.

Competence and Transformation

Within the biofilm, a fraction of *B. subtilis* cells can enter a state called **competence** once the population detects that it has reached a certain population density. This detection process is called **quorum sensing**: the population can gauge its cell density based on the concentration of pheromones secreted by the population.[83]

A cell exhibiting competence stops growth and becomes able to transport extracellular DNA into its own cell, and can incorporate this genetic information into its own genome. This is called **transformation**, and is an example of horizontal gene transfer (by contrast, vertical gene transfer happens through reproduction and inheritance). For *B. subtilis,* this state is also called **natural competence** since the bacterium can achieve competence with its own cellular machinery. In artificial genetic engineering, cells that lack natural competence need to be shocked with heat or chemicals to force them to absorb DNA, called artificial competence. Natural competence makes *B. subtilis* amenable to genetic modification.

Not all cells in a densely populated colony will display competence. Instead, the state of competence is controlled by a **stochastic switch**. To simplify, *B. subtilis* cells contain a genetic switch that is randomly turned 'on', causing a small number of cells within a growing population to enter the competent state. This random process is a way of 'bet hedging': competence can increase a population's chances for survival since it allows bacteria to horizontally transfer genes, which might be essential for survival. However, since the state of competence stops growth, it's advantageous if most of the cells are not competent and continue growing.[83]

Competence can be regarded as a crude form of sexual reproduction since it allows the shuffling of genetic information between cells. However, this isn't the only use of competence: external DNA can also be absorbed to repair existing genes, or simply be broken down as a nutrient.

Competence also plays a role in γ-PGA producing strains of *B. subtilis.* The same pheromones that are used to sense for cell density and competence are also used to induce the production of γ-PGA in cells growing on natto.[84]

The Multicellular Life Cycle

These mechanisms and tools give us a broader picture of the multicellular *B. subtilis* life cycle when placed in a batch culture, such as a container of natto, a petri dish, or a standing liquid culture. The life cycle can be summarized as follows.

During inoculation, a suspension of spores is introduced to a culture. In the subsequent lag phase, the spores will take some time to recognize that they are in favourable conditions, and eventually germinate, forming an initial population of vegetative *B. subtilis* cells.

With plenty of space and nutrients, these cells will rapidly reproduce and colonize the substrate during the exponential growth phase. In liquid cultures, individual cells will disperse through swimming motility and eventually form a pellicle on the culture surface. In solid cultures, cells will disperse along the surface using swarming and sliding motility, eventually thickening to form a biofilm. Strains without flagella, such as those used for modern natto production, will only be able to disperse through sliding motility, natural currents, agitation, or other means.

With plenty of food and nutrients, the biofilm will develop extracellular substances, fruiting bodies, and some cells of the growing population will enter a state of competence. As the biofilm matures, the supply of nutrients will eventually be exhausted and the rate of growth will equal the rate of death, and the colony enters the stationary phase.

Finally, in the death phase, the exhaustion of nutrients means that the rate of death will surpass the rate of growth. Under starvation conditions, the extracellular substances of the biofilm will be metabolized for growth. More and more cells will sporulate under these conditions, some of which will kill other cells and release more nutrients into the biofilm, providing the last reserves of nutrients for any remaining vegetative cells.

If adverse conditions continue, nutrients are depleted and the sample dries out, and eventually all that is left is a collection of endospores, which will remain dormant until they find favourable conditions again.

4. Strains of *Bacillus subtilis* and its relevance to Natto

The behaviours described in Section 3.2 are typically found in so called 'wild type' strains of *B. subtilis*, which exhibit a wide range of abilities, allowing them to adapt to different environments.

Domesticated strains like *B. subtilis* 168 (a commonly laboratory strain) and *B. subtilis* BEST195 (a common natto strain) have lost some of the 'wild-type' strain's abilities, since they were adapted to grow on clean, predictable substrates like a shake-flask culture or a package of steamed beans. In this section, we'll look at a few different strains of *B. subtilis* and how laboratory strains and 'wild-type' strains compare to natto-producing strains.

4.1 Strains of *Bacillus subtilis*

Unicellular Studies and *Bacillus subtilis 168*

For a long time, researchers regarded *B. subtilis* and many other bacteria as **planktonic**, treating bacteria as single-celled organisms that minded their own business and had a simple life cycle: if conditions were good, they would grow; and if not, they would sporulate. It wasn't until 1988 that the biologist James Alan Shapiro of the University of Chicago proposed that bacteria should be regarded as multicellular organisms, with more complicated behaviours.[85] These include intercellular communications and control systems, and the formation of specialized structures within a colony.

Part of the reason multicellularity was overlooked for so long in *B. subtilis* is that most of the research conducted on *B. subtilis* was performed on the laboratory strain *B. subtilis* 168.[86]

B. subtilis 168 is derived from a wild strain of *B. subtilis* named *B. subtilis* Marburg, after the German city from which it was isolated. Two Yale University botanists subjected *B. subtilis* Marburg to sublethal doses of Ultraviolet (UV) rays or X-rays, and found that many of the survivors became single-nutrient **auxotrophs**: the induced genetic mutations meant these bacteria were unable to synthesize specific compounds for survival (the original strain is called a **prototroph**). *B. subtilis* 168 is one of the strains derived from these experiments and was a tryptophan auxotroph: it could not synthesize the amino acid tryptophan, and can only grow in media containing this amino acid.

Auxotrophy is an important trait in genetic engineering techniques. For example, you can introduce a plasmid containing a novel gene into a culture of *B. subtilis* 168. If that plasmid also contains the gene needed to synthesize tryptophan, you can test if the bacteria successfully incorporated the plasmid by seeing if its ability to grow on media without tryptophan is restored.

The problem with studying *B. subtilis* 168 and other lab strains is that the mutation process also tends to disrupt many of the genes associated with multicellularity. Lab strains are also often grown in shake-flask cultures, where a flask of nutrient media is inoculated with an organism, shaken, then liquid samples are plated to obtain individual colonies. This process favours cells and colonies that are easily dispersed; cells that have lost most of their multicellular characteristics.

Multicellular Studies and *Bacillus subtilis* 3610

Studying wild strains of *B. subtilis*, rather than domesticated lab strains, revealed insights into how it behaves in a natural environment. Many of the behaviours described in the previous sections, including swarming motility, the production of surfactin, and the creation of aerial structures, are all found in wild strains of *B. subtilis* but not the mutated, domesticated *B. subtilis* 168.

In fact, most of the studies in multicellularity are somewhat recent, and required a paradigm shift starting with Shapiro's work in 1988 and continuing to present day.

Wild-type strains of *B. subtilis* are best exemplified by the strain *B. subtilis* 3610 (from NCBI3610, its identification number in the National Center for Biotechnology Information), which is also called the Marburg strain, and represents the original wild bacterium from which 168 was derived.

Not all strains of *B. subtilis* isolated from the natural environment exhibit the same set of traits as *B. subtilis* 3610. Within the species *B. subtilis*, there is a wide range of unique traits that leads to many varieties and subspecies, including traits that create the stickiness and aroma of natto.

Natto-Producing Strains and *Bacillus subtilis* BEST195

Natto-producing strains of *B. subtilis* were artificially selected and domesticated to produce good natto. This means selecting for strains that produce the most mucilage through large amounts of the EPS, poly-γ-glutamic acid (γ-PGA).

Functioning flagella inhibit the synthesis of γ-PGA, so modern natto-producing strains of *B. subtilis* aren't able to swim, and instead produce large amounts of γ-PGA.[87] This is not a problem for the bacterium, since natto spores are introduced in large numbers to mostly sterile, solid environments of steamed beans, and have little to no competition. Before the introduction of pure cultures, both flagellated and non-flagellated *B. subtilis* strains were identified in early experiments with rice-straw natto (see Chapter I.4.1).

Miyagino, Takahashi, and Naruse are the three major natto starter producers in Japan (see Chapter II). In academic literature, the Miyagino strain is used as the reference genome under the strain name *Bacillus subtilis* BEST195, although the three strains are nearly identical.[88,89] These strains are synonymous with the designation *B. subtilis* var *natto.*

Besides the plentiful production of γ-PGA and the lack of flagella, the other notable property possessed by BEST195 is that it is a biotin auxotroph (i.e. requires biotin for growth), which is not a problem since the vitamin biotin is found in sufficient quantities in steamed soybeans.

BEST195 also possesses many copies of **insertion sequences**, which are short sequences of DNA that can 'jump' positions within a genome. One of these insertion sequences, IS*4Bsu1*, can insert itself into the *comP* gene which regulates the *pgs* operon responsible for producing γ-PGA. If this happens, BEST195 can spontaneously be rendered unable to produce γ-PGA. In fact, by repeatedly re-plating BEST195, one can breed a strain of *B. subtilis* var *natto* that does not produce any γ-PGA, allowing for the creation of 'non-sticky' natto. This exact strategy was used to develop the non-sticky brand of natto '*Mamenoka*' without any specialized genetic engineering or mutative procedures. The development of *Mamenoka* will be detailed later.

Biotin auxotrophy and the presence of insertion sequences are common traits of commercial natto strains, but they are not strictly required for natto production. This raises a fundamental question: what exactly constitutes a 'natto-producing' strain? Historical natto literature would suggest that the two basic requirements are 1) the production of a sticky substance and 2) the 'aroma' of natto. However, the aroma of natto is subjective, and stickiness is also not a requirement if 'non-sticky' natto is still considered natto. Thus, the variety *B. subtilis* var *natto* isn't clearly defined either, which returns us to the problem of bacterial taxonomy.

One could argue that any bacterium, not just *B. subtilis*, could produce a 'natto', so long as it has the desired properties of taste, texture, aroma, and safety. Furthermore, you could speculate that a mixture of different species–not just a monoculture of *B. subtilis*–could be used to create natto. After all, the original experiments isolating bacteria from straw natto in the early 20th century yielded multiple species of bacteria, including *cocci* and both flagellated and non-flagellated strains of *B. subtilis.*

The criteria for what strain of *B. subtilis* makes 'good natto' is subjective and up to the consumer. In the future, we may see further developments in natto-producing strains, targeting different aromas, tastes, and functional ingredients.

4.2 Subtilisin and Nattokinase

The question of what constitutes 'natto-making' strains of *B. subtilis* is complicated even further by the discovery of **nattokinase** by Hiroyuki Sumi in 1980, with a paper published in 1987. Sumi discovered that one of the **proteases** (protein-degrading enzymes) secreted by *B subtilis* var. *natto* is effective at breaking down fibrin, one of the proteins involved in human blood clotting.[90] When ingested, this can lead to physiological effects that include lower blood pressure, and the reduced risk of cardiovascular diseases. These effects will be explored in detail in Chapter VI.

Nattokinase is simply a version of the extracellular enzyme **subtilisin**, which is secreted by the *B. subtilis* into the environment to break down proteins for consumption. In *B. subtilis* 168 and other common strains, subtilisin is coded by the gene *AprE*, and the enzyme is known as 'subtilisin E'. This enzyme is incredibly useful in industry, where *B. subtilis* or *B. amyloliquefaciens* is grown and the subtilisin is harvested for uses in food processing, detergents, cosmetics and cleaners.[91]

However, in the natto-producing strain *B. subtilis* BEST195, a few substitutions in the genetic code change the subtilisin ever so slightly. This version of subtilisin can still degrade proteins, but also happens to be effective at degrading fibrin. In BEST195, this form of subtilisin is known as 'subtilisin NAT', and is coded by the gene *AprN* which is 99.2% identical to *AprE*. Of the 1146 base pairs making up this gene, only 9 are different between *AprN* and *AprE*.[92] Sumi coined the term nattokinase in reference to the human clot-dissolving enzyme **urokinase**, but technically nattokinase is a subtilisin rather than a kinase.

It's happenstance that BEST195 produces the fibrin-selective protease AprN, but this isn't a rare characteristic for microbial proteases. A whole host of other *B. subtilis* strains, along with other microorganisms, were found to secrete proteases that also had strong fibrinolytic activity.[93] Fibrin is a protein, so in general it can be broken down by proteases, but some proteases are better at degrading fibrin than others. This is the nature of protein-to-protein interactions, where some proteins simply 'fit' better with others.

The health benefits of nattokinase may be important for natto's role as a functional food, and by extension, its marketing. This raises the question: what is the full potential of natto as a functional food? If we examine natto from a purely nutritional and functional perspective, rather than just its organoleptic (taste, smell, visual, and textural) characteristics, is it possible to develop a natto product that maximizes its health benefits?

Recall that many researchers missed out on many of the multicellular behaviours of *B. subtilis* because of their focus on one single strain, *B. subtilis* 168. Similarly, I hope natto producers will not limit their potential by only using *B. subtilis* BEST195 and its derivatives, and continue to explore novel and newly isolated strains of *B. subtilis.*

5. Molecular Biology of *Bacillus subtilis*

At the heart of every cellular function described thus far is a system of molecular machinery encoded by the *B. subtilis* genome. Bacteria were historically dismissed as a simple "bag of enzymes", using molecular diffusion to achieve all of its cellular functions.[94] Modern research has since revealed that even within such seemingly simple microorganisms, highly complex mechanisms can be found. Describing all these mechanisms is far beyond the scope of a natto textbook. However, we can examine some of the basic concepts of molecular biology in *B. subtilis*, including its genome, and how genetics affect the final natto product. These concepts might help distinguish between 'natto-producing' and 'non-natto producing' variants of *B. subtilis*. Genetically modified or artificially selected strains of *B. subtilis* have also been used to produce new, commercially successful natto products.

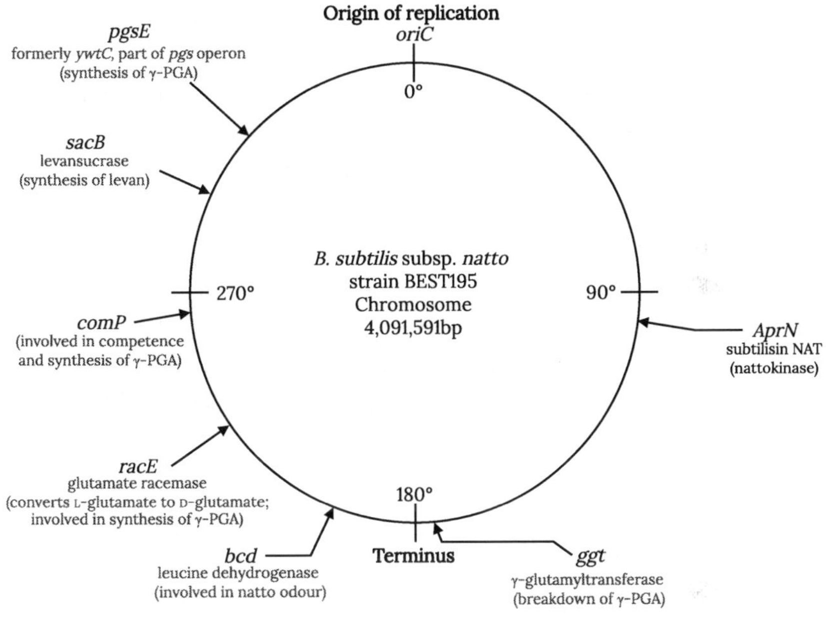

Figure IV.11 Bacillus subtilis strain BEST195 showing selected genes

5.1 The Bacillus subtilis Genome

Like all bacteria, *B. subtilis* is **haploid**, and contains a single copy of its chromosome unless reproducing. This chromosome is organized in a ring of approximately 4.1–4.2 million base pairs (Mbp).

Molecular biologists organize the chromosome around the **replication origin,** denoted ***OriC*,** which is given the 0° position. As the cell grows, DNA replication starts at *OriC* and two replication forks proceed bidirectionally in the clockwise and counterclockwise direction, with each fork reaching the 90°

and 270° positions, then finally meeting at the 180° position (the terminus), after which the chromosome copies separate.[95]

In a growing cell, the chromosome is replicated before cell division initiates, so *B. subtilis* cells can be found with two full chromosome copies late in its cell life, with the 0° position *OriC* located near the poles and the 180° position located near the cell centre. During binary fission, the cell wall forms in the middle of the cell and separates these two chromosomes, yielding two daughter cells each with a copy of the original.

During sporulation, the cell wall begins to form, but the smaller forespore has its chromosome pinched off by the cell wall. In this case, the chromosome is rescued by an enzyme called DNA translocase, which 'pumps' the cut-off portion of the chromosome back into the forespore. This can also occur if one of the two chromosomes is accidentally cut off during regular binary fission.[76]

DNA translocase is just one of many molecular machines that *B. subtilis* can use to ensure successful replication and the preservation of its genome. There are also other mechanisms that repair broken DNA, and edit any mistakes made during chromosome replication.

The chromosome itself doesn't form an actual circular ring within the bacterium, otherwise this ring's diameter would be on the order of 1000x the bacteria's length. Instead, the chromosome is clumped up in the centre of the cell in a densely packed region called the nucleoid, from which loops of DNA protrude outwards.

Some strains of *B. subtilis* can also contain **plasmids**, which are smaller circles of DNA separate from, and much smaller than the main chromosome. These plasmids only contain a handful of genes and replicate independently from the main chromosome. Plasmids are sometimes useful when introducing new traits to a bacterium, and they can confer an evolutionary advantage.

The Nucleoid and Ribosomes

An abridged form of molecular biology is as follows. DNA replicates itself, enabling reproduction and heredity. RNA polymerase **transcribes** individual genes within DNA into messenger RNA (**mRNA**). mRNA, with the help of **ribosomes** and transfer RNA (**tRNA**), **translate** the mRNA into proteins. These proteins carry out the functions of the cell, including metabolic and structural uses, among many others.

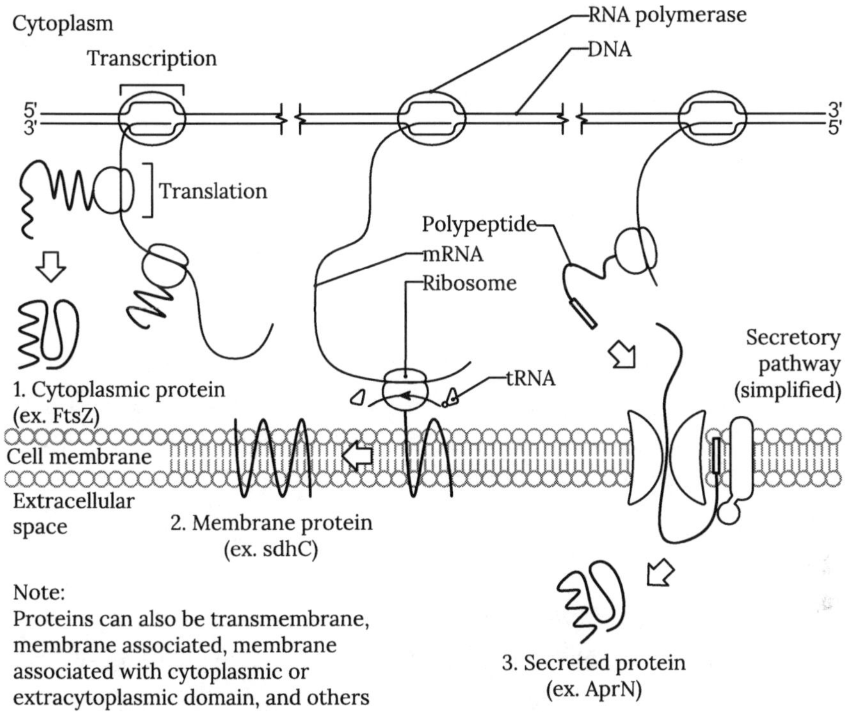

Figure IV.12 Transcription and translation

The complete set of mechanisms is much more complex. Surrounding the processes of transcription and translation is an entire system that regulates gene expression, which can involve proteins, different types of RNA, and signalling molecules that interact with these components.

In *B. subtilis* and other bacteria, there isn't a nucleus that physically separates transcription and translation. The two processes can happen simultaneously, with multiple ribosomes translating a strand of mRNA as it is being transcribed from DNA. This allows the translation of many proteins in parallel by several ribosomes, which can also protect the fragile strand of mRNA as it's being processed.[96]

Individual ribosomes (size 70S) are composed of large (50S) and small (30S) subunits, and are often depicted in microbiological cartoons as a sort of burger bun. The 'S' used to describe the ribosome and its subunits refers to **Svedburg units**, which is a measure of how quickly a particle moves when in a centrifuge, or its settling velocity. Svedburg units can be roughly interpreted as its size and becomes relevant in the discussion of soybean proteins in the next chapter. Each ribosomal subunit is composed of a mixture of proteins and rRNA bound together. For example, the 30S small subunit is composed of 16S rRNA and ribosomal proteins. The genetic code for 16S rRNA is commonly used in bacterial phylogeny (see Section 2).

These ribosomes are too big to diffuse into the dense nucleoid at the centre of the bacterial cell, so they're generally found in the cytosol closer to the cell membrane. Proteins that are destined for the bacterial membrane can anchor the ribosome to the membrane. The bacterial chromosome can form 'loops' that extend outwards towards the membrane and ribosomes. Genes located deeper within the nucleoid can still be transcribed, but the transcribed mRNA has to be 'shuttled' out of the nucleoid through a variety of chaperone proteins that bring it out of the nucleoid.[96]

Insertion Sequences

Certain strains of *B. subtilis*, including BEST195, possess sequences of DNA known as **insertion sequences** (IS). These sequences, known informally as 'jumping genes', can transpose themselves into other parts of the chromosome, sometimes disrupting other genes.

In BEST195, the first IS was discovered in 2000 and named IS*4Bsu1*. This IS can cut itself from one location, then paste itself elsewhere in the chromosome.[97] This occurs preferentially at certain sites within the chromosome, including the gene *comP*, which is involved in competence, and ultimately stops the bacterium's ability to produce γ-PGA. Before this discovery, it was known to natto researchers that BEST195 would naturally produce mutants that were defective in producing γ-PGA at a rate of about 1% in every 20 generations.[98] The discovery of IS*4Bsu1* provided an explanation to this random defect.

Interestingly, when Japanese researchers went to collect rice straw samples from across Japan to isolate wild strains of *B. subtilis*, they found that the majority (33 out of 59) of natto-producing strains possessed IS*4Bsu1* along with another insertion sequence, IS*256Bsu1*, indicating that this is a common trait. The other 26 of the 59 strains did not contain either IS. These strains have the potential of being used as natto starter in the future, and would not be vulnerable to spontaneous mutations that render them unable to produce γ-PGA.

Table IV.2 Properties of the B. subtilis genome

Main Chromosome Property Note 1	***B. subtilis* BEST195**	***B. subtilis* 168**	***B. subtilis* 3610**
Base pairs in chromosome	4,105,390 Note 3	4,215,606	4,214,598
Genes (total)	4,375	4,536	4,543
Protein coding genes	4,037	4,237	4,230
Genes (RNA)	122	118 Note 2	122
Ribosomal RNA (rRNA) coding genes	10, 10, 10 (5S, 16S, 23S)	10, 10, 10 (5S, 16S, 23S)	10, 10, 10 (5S, 16S, 23S)
Transfer RNA (tRNA) coding genes	87	86	87
Non-coding RNA (ncRNA) genes	5	2	5
NCBI Ref. Sequence	NC_017196.2	NC_000964.3	NZ_CM000488.1
Plasmids	**2**	**0**	**1**

Note 1: The most updated sequences as of Dec. 2022 are presented here; note that the exact number of genes, base pairs, and their function may change with new research.
Note 2: In addition to these 118 RNA genes, this sequence also contains 93 identified miscellaneous RNA sequences that include regulatory RNA and riboswitches.
Note 3: This is slightly different than in Figure IV.11 due to different sequencing results

5.2 Key Metabolic Pathways

Basic Metabolism

B. subtilis can metabolize common sugars, alcohols, proteins, and fats into energy. The act of breaking down these substances is called **catabolism**. Compounds can follow multiple routes as they are broken down or converted into different molecules, and each step is catalyzed by an enzyme. Intermediate and by-product compounds are called **metabolites**, and many of these metabolites contribute to the unique flavour and aroma of natto. Intermediate compounds can also be used to build other molecules used by the cell for growth, a process called **anabolism**. Collectively, the conversion steps from one compound to another is called a **metabolic pathway**.

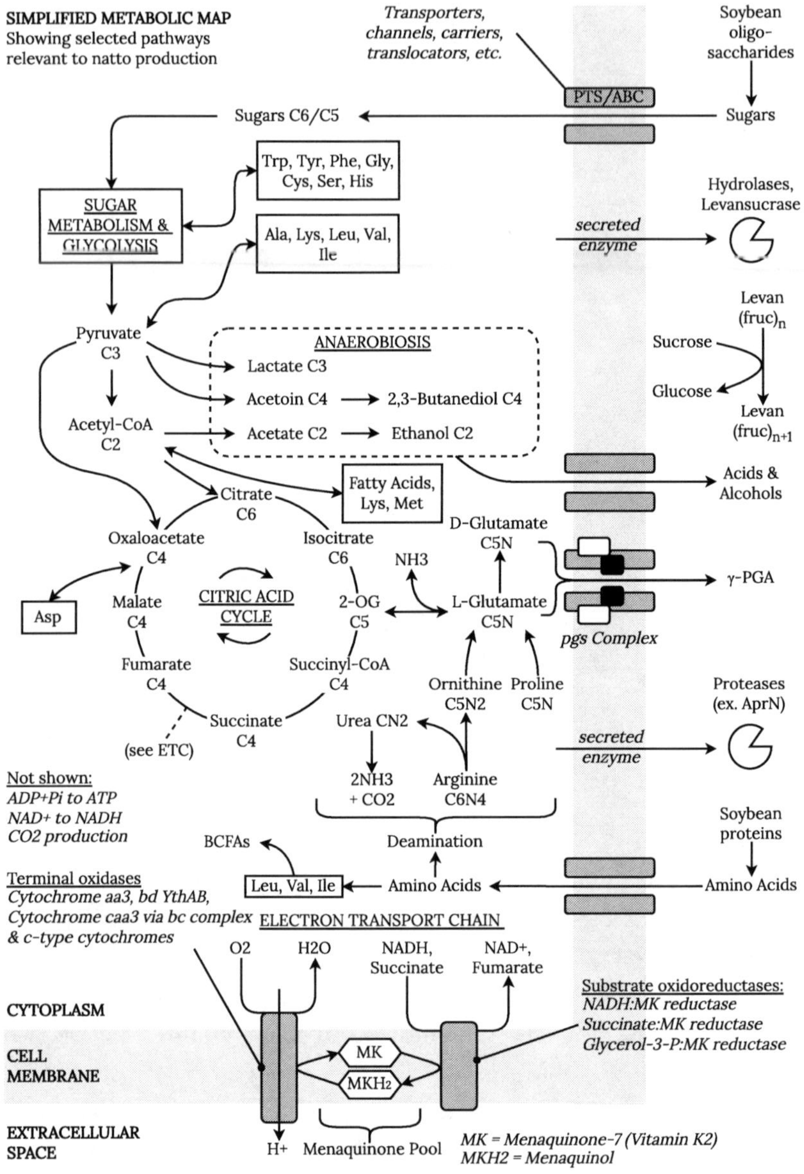

Figure IV.13 Map of B. subtilis metabolic pathways, esp. pertaining to natto. Not all pathways and mechanisms are shown, for ease-of-understanding. For a full metabolic map, see Further Reading. Most metabolites show only Carbon and Nitrogen numbers.

Glycolysis and Related Pathways

In the glycolytic pathway, 6–carbon glucose is broken down into 3–carbon pyruvate. Other sugars such as mannitol, mannose, fructose, and alcohols such as glycerol are connected to this pathway at intermediate points, as well as some of the amino acids. Pyruvate itself can also be converted into several other amino acids, lactate, and acetoin, the latter of which is a major aroma component of natto.[99]

Acetyl–CoA and the TCA Cycle

3–carbon pyruvate is further broken down in Acetyl-CoA, which consists of a 2–carbon acetyl group bonded to Coenzyme-A. Acetyl-CoA is often considered the 'centre' of a metabolic map, and sits at the crossroads of many other pathways. Acetyl–CoA can be used to synthesize several amino acids and fatty acids. It also feeds the tricarboxylic acid (TCA) cycle, also known as the citric acid cycle or Kreb's cycle.

The TCA cycle extracts the chemical energy out of the 2–carbon acetyl group in Acetyl–CoA during aerobic growth. The cycle starts by joining the 2–carbon acetyl group to 4–carbon oxaloacetate to form 6–carbon citrate, which goes around the TCA cycle in successive steps, ultimately reforming oxaloacetate to restart the cycle. The whole process extracts energy in the form of adenosine triphosphate (ATP) and other intermediates, yielding 2 molecules of carbon dioxide.[100]

Electron Transport Chain and Menaquinone (Vitamin K_2)

Catabolic processes generally yield ATP from adenosine diphosphate and a phosphate group (ADP + P_i), which can be thought of as an 'energy currency' that can directly for cellular processes requiring energy.

Catabolic processes also yield an energy intermediate, nicotinamide adenine dinucleotide (NAD) in its reduced form, NADH, from its oxidized form, NAD^+. The final step in respiration is a series of steps that take place around the cell membrane and ultimately reduces molecular oxygen, O_2, into water, while oxidizing NADH back into NAD^+ for reuse as an electron carrier intermediate.

These steps, known as the electron transport chain (ETC), successively transfer electrons from one carrier to another, and the overall ETC has a net effect of pumping hydrogen ions (H^+) outside of the cell membrane. This produces a H^+ gradient between the inside of the cell membrane and outside (between the cell membrane and the cell wall). The electrical potential of this H^+ gradient, which acts like a capacitor or battery, is harnessed to generate additional ATP from ADP + P_i using ATP synthase. This electrical potential is also used to drive other processes around the cell membrane, including rotation of the flagella, active transport of substances across the membrane, and other functions.[101]

In *B. subtilis,* the ETC is broken down into three classes of components. The first component is a collection of membrane bound protein complexes known as **substrate oxidoreductases.** These complexes oxidize a substrate back into its oxidized form and reduces membrane-bound **menaquinone** into **menaquinol.** Example substrates are NADH, which is oxidized into NAD^+, and succinate, which is oxidized into fumarate (this reaction is part of the TCA).

Menaquinone and menaquinol are lipid soluble molecules, and reside within the cell membrane, allowing them to move freely between enzymes and other membrane-bound protein complexes. Together, they form the second component of the ETC known as the **menaquinone pool.** Menaquinone is also an essential vitamin for humans, Vitamin K_2.

The third component of the ETC is another collection of membrane bound complexes known as **terminal oxidases.** These come in variety of forms, and they oxidize menaquinol back into menaquinone, transferring its electrons to molecular oxygen O_2 to yield water. This process pumps H^+ ions outside the cell membrane, completing the ETC.

Different organisms have different configurations, molecules and enzymes used within the ETC. The fact that *B. subtilis* (and many other aerobic, gram-positive bacteria) uses menaquinone as an intermediate electron carrier is the whole reason natto happens to be such a rich source of Vitamin K_2.

Anaerobiosis and Anoxic Conditions

In anaerobic conditions, *B. subtilis* cannot use the ETC to regenerate NAD+ from NADH. Instead, it will make use of several fermentative pathways to regenerate NAD+. These pathways include the conversion of acetoin into 2,3–Butanediol, pyruvate to lactate, acetyl–CoA to acetaldehyde, and acetaldehyde to ethanol. These pathways occur in limited amounts during normal natto production, but the overabundance of acids and alcohols could indicate insufficient oxygen.[66]

B. subtilis can also use nitrate (NO_3^-) or nitrite (NO_2^-) as an electron acceptor in the absence of oxygen. Nitrate is reduced to ammonium via nitrite, while nitrite reduction oxidizes NADH into NAD+.[66] These pathways are not expected in natto production, and only occur in water-logged soils or marine environments.

Biofilm Chemistry: Polyglutamic Acid, Levan and TasA

Natto-producing strains of *B. subtilis* growing on sucrose and glutamate-rich media will form a biofilm that is rich in levan, a polymer of fructose, and γ-PGA, a copolymer of L– and D–glutamate.

L–glutamate and D–glutamate are stereoisomers and can be visualized as left-handed and right-handed mirror images of each other. In biochemistry, this makes them functionally different. L–glutamate is the form used to synthesize proteins, but D–glutamate can be synthesized by *B. subtilis* through glutamic acid racemase and D–amino acid aminotransferase. Natto-producing strains of *B. subtilis* use both stereoisomers of glutamate when producing the polymer γ-PGA, which is sometimes referred to as γ-DL-PGA. This polymer of glutamic acids is not formed by ribosomes, but a membrane-bound enzyme complex (the pgs complex or PGA synthase complex) that can form molecules of γ-PGA 10–1000kDA in size.[102,103]

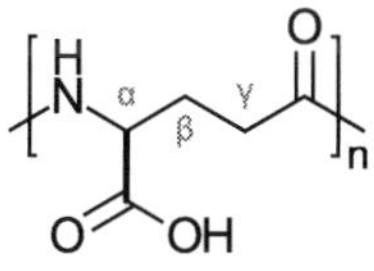

poly-γ-glutamic acid

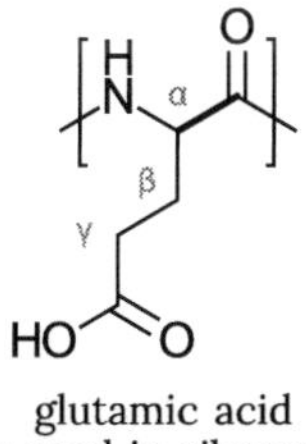

glutamic acid
(as used in ribosomal
protein synthesis)

Figure IV.14 Comparison of γ-PGA and protein glutamic acid

The prefix 'γ' indicates these L– and D– glutamate monomers are linked together by an amide ligation system, which bonds the amine group of one monomer to the γ carbon of another monomer. This is quite different than proteins, where the amine group of one amino acid bonds to the α carbon. As a result, γ-PGA is structurally different from α-PGA, and the former is resistant to digestion by proteases. In later stages of growth, *B. subtilis* secretes an enzyme called gamma-glutamyl transpeptidase (GGT) to break gamma-PGA back into its constituent monomers.[104]

When growing in the presence of sucrose (a disaccharide of glucose and fructose), *B. subtilis* will create levan as a component of its biofilm. Levan is a polymer of fructose formed by β-(2→6) linkages with occasional β-(2→1) branching. This is catalyzed by the enzyme levansucrase, which breaks down sucrose into glucose and fructose, adding the fructose to the growing levan polymer, while liberating glucose. Like γ-PGA, levan can be leveraged as a store of energy and be broken down into fructose as energy stores are depleted.[105]

The final major component of the *B. subtilis* biofilm is a protein called TasA, which forms amyloid-like fibres within the biofilm. These fibres bind cells together within the biofilm, bind cells to substrate surfaces, and provide structural integrity to the biofilm matrix.[106]

Collectively, this biofilm forms the 'mucilage' that surrounds the soybeans in natto. Of all the biofilm substances, γ-PGA is the substance primarily responsible for natto's viscosity. Maximizing stickiness depends on optimizing the growth of *B. subtilis*, since insufficient fermentation will yield only a small amount of γ-PGA, while overfermentation will cause the bacteria to break down its own γ-PGA for food.

Subtilisin and Ammonia-Forming Reactions

As free sugars are depleted from the bacterium's environment, *B. subtilis* will rely more and more on amino acids as a carbon and energy source. This happens as the bacterial population reaches latter stages of growth.

Firstly, *B. subtilis* adapts by secreting more proteolytic enzymes into its environment to break down proteins. In BEST195, the major proteolytic enzyme is AprN (subtilisin NAT, also known as nattokinase) and in 168, this enzyme is AprE (subtilisin E). These secreted enzymes break down proteins into amino acids, which can diffuse back to the cell.

Since amino acids contain nitrogen, they must undergo a reaction that removes the amine group before they can be used as a carbon source. In general, *B. subtilis* will prefer to use the glutamate family of amino acids (glutamate, glutamine, arginine, proline) and aspartate (asparagine) for this purpose.[107]

One major source of ammonia in natto production is the breakdown of glutamate into α-ketoglutarate (also called 2–oxoglutarate, or 2–OG) and ammonia (NH_3) through the enzyme glutamate dehydrogenase. 2–OG can then be used as an energy source via the TCA cycle.[107] Other amino acids can be converted into glutamate before undergoing this reaction, namely proline (which generates no additional ammonia) and arginine. The conversion of arginine into ornithine generates urea as a by product. Urea is then converted by the enzyme urease into two molecules of ammonia.

Ammonia formation is unavoidable in natto production since these pathways are always active. However, ammonia production can be controlled by limiting the fermentation time and ensuring that *B. subtilis* has plenty of free sugar to consume.

Summary of Metabolism and its Relevance to Natto

There are many other aspects of *B. subtilis* metabolism that are not fully described in this section. However, the key points that are relevant to natto are as follows:

1) *B. subtilis* prefers sugars (glucose, fructose, mannose, raffinose family sugars) as its carbon source, but can use amino acids as well, all of which feed into multiple points within the bacterium's metabolic pathways.
2) *B. subtilis*, and other gram-positive aerobic bacteria, use Menaquinone as an intermediate electron carrier within its electron transport chain. This is why natto is such a rich source of Vitamin K_2.
3) In the presence of excess sucrose and glutamate, *B. subtilis* can synthesize levan and poly-γ-glutamic acid, which are two main constituents of the mucilage in natto. However, as sugars and amino acids become exhausted, *B. subtilis* will start to break down this mucilage as a store of energy.
4) As sugars become exhausted, *B. subtilis* will also begin to use more and more amino acids for energy. This process requires deamination, which ultimately releases ammonia as a by-product.

6. Microbial Techniques and Natto Starter

With the fundamentals of *Bacillus subtilis* in hand, let's take a look at the applied microbiology techniques used to develop spore solutions (*natto-kin*), breed different strains, and isolate novel strains from nature. These techniques have been detailed by many researchers working with natto and *B. subtilis*.

Caution should be exercised when working with bacteria isolated from the natural environment, as there are a few deadly pathogens including *Bacillus anthracis, Bacillus cereus,* and *Clostridium difficile* that are also endospore-forming bacteria found in soil environments. Unless you absolutely know what you're doing, this section should be for knowledge only. You should also know what commercial natto looks, smells, and tastes like before consuming any homebrewed natto, especially with bacterium isolated from nature.

6.1 Isolation of Natto-Producing Bacteria from Nature

One method for isolating 'wild' strains of *B. subtilis* for natto comes from a 2011 Japanese research paper, where researchers were trying to answer fundamental questions, such as whether biotin auxotrophy and insertion sequences are always found in natto-producing strains. These researchers scoured all of Japan, from Okinawa to Hokkaido, and collected a total of 104 rice straw samples across 14 areas, with most of them were harvested from Ibaraki prefecture. From these random samples, they isolated bacteria from rice straw using the following screening steps.[89]

1. Samples were heated in sterile water at 95°C for 5 mins. This would remove any vegetative cells from the straw and only spores would remain.
2. An equal volume of tryptic soy broth (TSB) was added to the water and the sample was allowed to sit overnight at room temperature to germinate the spores.
3. Samples were diluted, then plated onto TSB agar plates to grow colonies.
4. Individual colonies were transferred to glutamate starch phenol (GSP) plates to assess γ-PGA synthesis. Non-γ-PGA producing strains were removed.

5. Each strain was grown on soybeans using a standard procedure to create individual natto samples (see Chapter II).
    a. Small soybeans were soaked for 16h at 20°C.
    b. Soaked soybeans were steamed at 131–133°C for 30 minutes.
    c. Steamed soybeans were fermented at 39°C, for 18h at 90% RH.
    d. The temperature dropped to 20°C, for 2h at 50% RH.
    e. Natto samples were matured at 5°C, for 24h.

Note that in laboratory settings, a small scale automatic natto fermenter (model SY No. 20 by Suzuyo Kogyo Ltd.) is often used, which comes with programmable temperature and humidity.

Individual natto samples were then evaluated to determine its quality (see Chapter III). Strains could be deemed unfit for natto production based on any of the following:

1. Non-uniform or non-homogenous growth.
2. Formation of a syrup-like biofilm.
3. Insufficient fermentation (product was too hard).
4. Lack of a sticky texture.
5. Strong smell of ammonia.

From the 104 rice straw samples, 424 spore-forming, γ-PGA producing samples were isolated for testing on steamed soybeans. These were dominated by *B. subtilis* and the closely related *B. amyloliquefaciens*, although other strains such as *B. pumilus*, *B. atrophaeus*, and *B. licheniformis* were found.

From these 424 strains grown on steamed soybeans, only 59 strains remained after the quality screening process, all of which were *B. subtilis*. Interestingly, every single one of them were found to be biotin auxotrophs. Of these, 26 did not possess the either IS element IS*4Bsu1* or IS*256Bsu1*, while 33 strains possessed both.

The experiment suggests that biotin auxotrophy is highly correlated with the ability to produce good natto, although the exact reason for this is yet to be elucidated. On the other hand, IS elements are common, but expectedly not required for natto production.

The fact that *B. amyloliquefaciens* was ubiquitous in rice straw samples begs the question as to whether this bacterium was commonly found in Edo period natto. From phylogenetics, we know that *B. amyloliquefaciens* is closely related to *B. subtilis*, with the primary difference that it produces plenty of α-amylase, which can hydrolyze starch into sugar, characterized by a syrupy biofilm.

The bacteria used to evaluate natto production were screened as being positive for γ-PGA production. Strains of *Bacillus subtilis* that are negative for γ-PGA production should still be considered since there is a demonstrable market demand for non-sticky natto. Processed natto products, like dried natto (*hoshi natto*, see Chapter III.2.2) also benefit from having low stickiness for ease of processing.

The same screening technique described above can be modified to isolate other pure *Bacillus* strains from nature for natto production. The key steps for the isolation of *B. subtilis* from nature can be summarized as follows.

1. Heat (boil) rice straw to isolate sporulated bacteria, then germinate the spores on a liquid culture (e.g. TSP broth)
2. Culture the germinated spores on petri dishes (e.g. TSP agar) to isolate individual colonies of a single strain.
3. Re-grow the culture on GSP media to assess γ-PGA synthesis (if required).
4. Create a 'test batch' of natto using the isolated culture (see methods in Chapter II).
5. Discard any batches of natto not meeting the user's sensory criteria (see methods in Chapter III).

A Note on Growth Medium

Users should employ the appropriate growth media depending on the culture objectives. In general, **undefined media** containing complex hydrolysates like beef peptone and soy peptone are appropriate for general growth since they contain all the compounds needed by the bacteria. These include tryptic soy broth (TSP), natrium beef peptone (NBP), Difco nutrient broth, and others. Glutamate starch phenol (GSP) media is specifically used to assess γ-PGA synthesis, since the glutamic acid provides the glutamate monomers needed to synthesize γ-PGA. The addition of trace manganese helps with sporulation– this will be described later.

By contrast, **defined media** contains known quantities of all compounds, and can be used to test for characteristics like auxotrophy. For example, a defined medium without biotin can be used to screen for biotin auxotrophy.

6.2 Identification Methods

Once a strain of natto producing bacterium has been properly isolated, there are several techniques that can be used to confirm the identity of this microorganism.

Phenotypic Methods

Table IV.3 Selected characteristics of B. subtilis from Bergey's Manual.[73]

Characteristic	*B. subtilis*	*Notes on B. subtilis* var *natto*
Gram stain	Gram +ve	
Motility	+	*Typical natto strains non-motile*
Length	2.0–3.0μm	
Diameter	0.7–0.8μm	
Spore formation	Ellipsoidal-cylindrical	
Parasporal crystals	–	
Catalase	+	
Aerobic growth	+	
Anaerobic growth	–	
Voges-Proskauer Test	+	
Egg yolk reduction	–	
Nitrate reduction	+	
Optimum Growth Temp	28–30°C	38°C on steamed soybeans
Minimum Growth Temp	5–20°C	
Maximum Growth Temp	45–55°C	>55°C causes product failure
pH Range	5.5–8.5	
Salt tolerance	up to 7% NaCl some up to 10% NaCl	
Salt required for growth	–	

Refer to Bergey's manual for full list of differential characteristics.

Phenotypic methods rely on direct observation of the bacteria's traits. The most recent edition of *Bergey's Manual of Determinative Bacteriology,*[108] and *Bergey's Manual of Systematic Bateriology,*[73] should be used. These methods are considered dated compared to molecular methods, but are still useful for identifying bacteria, especially down to the genus. They are a good option available for users without access to DNA and molecular identification techniques.

Bergey's manual relies on using successively finer characteristics to narrow down the isolated bacteria's identity. The first identification technique is the gram stain, followed by observing the bacterium's shape. Finer details require observing the bacterium's pigment, motility, cell size, spore shape, oxygen utilization (aerobic vs anaerobic growth, reduction of nitrates), and the presence of enzymes such as catalase. The finest details involve testing the bacterium's growth on a variety of substrates.

Genotypic Methods

Advanced methods of identification rely on molecular techniques by sequencing the bacterium's DNA. Most of these molecular techniques require amplifying a DNA sequence using **polymerase chain reaction** (PCR), which works as follows:

1. Double-stranded DNA (dsDNA) is purified and separated into single stranded DNA (ssDNA) using heat.
2. Primers are annealed to the individual ssDNA strands.
3. dsDNA is synthesized from the strands of ssDNA using an enzyme, typically *Taq* polymerase, which is a heat-stable DNA polymerase from the thermophilic bacterium *Thermus aquaticus.*

Repeating these steps yields a large quantity of DNA amplified from the original culture, known as an **amplicon**. PCR is used in many related identifications techniques such as whole genome sequencing.

Whole genome sequencing (WGS) used to be prohibitively expensive, and only select organisms would be fully sequenced. Today, whole genomes can be sequenced inexpensively: the original human genome project cost US $2.7 billion and 10 years to come up with the first human genome sequence. Nowadays, the full human genome can be sequenced for less than $1000 USD and a few days on a **next-generation sequencing** (NGS) machine.[109]

For simple bacteria and microbes, sequencing by WGS using NGS can be completed in under 24h. NGS also allows for whole genome sequencing of numerous organisms in one sequence run, which also allows users to study **metagenomics**. Common platforms for NGS are made by Illumina,

ThermoFisher, Oxford Nanopore, and Pacific Biosciences.[110] The general steps in NGS are as follows.[111]

1. A bacterial culture is grown on agar plates.
2. The DNA is extracted by lysing the bacteria and purifying the DNA.
3. The DNA is sheared by enzymatic or mechanical means.
4. The DNA is amplified using PCR, forming an amplicon.
5. The amplicon is loaded into a sequencer, which determines the sequence of each fragment, known as a 'DNA read'. Each organism can produce millions of DNA reads.
6. The DNA reads are fed into software for DNA sequence analysis, which reconstructs the whole genome by comparing and overlapping DNA reads.

The resulting whole genome sequence can be used to determine the species through online databases. Other genome sequencing methods are available, but they are like WGS in methodology.

WGS using NGS is incredibly powerful, and can be used to identify the small differences needed to distinguish between individual strains within species.

6.3 Starter Preparation (Natto-kin)

Once a desirable strain of natto-producing *B. subtilis* has been isolated, a starter solution of the strain's endospores, or *natto-kin*, can be prepared for natto production.

Sporulation is induced by starving *B. subtilis* of nutrients. In the laboratory, the procedure for producing spores is also aided by using a sporulation media. This media must contain a small amount of manganese, which is essential for the sporulation process.[112] Many different forms of media designed for bacterial sporulation have been developed by life sciences companies.

It's not clear exactly what methods are used by the commercial natto spore producers Miyagino, Naruse, and Takahashi. In this section we rely on two methods from the technical book 'Investigative Approach to Natto', which references a method by Jiro Miura, who founded Miyagino Natto, and had been working with the production of natto starter starting in the 1950s.[113]

'Plate Scraping' Method

Materials:

- Liquid medium (NBP, Natrium Beef Peptone): 0.5% NaCl, 1% Beef Extract, 1% Phytone Peptone.
- For sporulation medium, add $MnSO_4$ at 0.1mM to above medium.
- For plate medium, add 2% agar to above medium.

Everything must be sterilized ahead of time according to laboratory best practices. The method is as follows:

1. Inoculate bacteria sample on plate medium and incubate overnight at 37°C (Note: 'overnight' is not defined. The purpose of this step is to germinate frozen bacterial stock or endospores).
2. Using an inoculation loop, isolate one colony and grow in 100mL of liquid medium, with shaking at 110rpm at 37°C for 24h.
3. Inoculate the cultured bacteria onto a plate sporulation medium and incubate at 37°C for 24h.
4. Scrape colonies off the above plate with a micro spatula, and suspend in 10mL of sterilized water. Vortex as appropriate to suspend the colonies.
5. Centrifuge at 12000rpm for 10 minutes. Suspend the sedimented cells in sterilized water and centrifuge again. Repeat twice, finally suspending the cells in 10mL of sterile water.
6. Store the suspension at 4°C for 24h.
7. Mix the 10mL of cell suspension with 25g of sea sand, and heat at 40°C for 40h.
8. Dry in a vacuum desiccator.
9. Add 10mL of sterilized water and stir well to make the starter suspension.

Note: in this method, spores tend to clump together. The sea sand added in step 7 is used primarily to break up the spore suspension.

Pellicle Method

Materials:

- Liquid medium: soybean meal, glucose, peptone.
- Note: the author does not give exact measurements of the medium ingredients, and references an original method from 1949 in

'Rational Manufacturing Method of Natto'.[114] The medium may be similar to the modern Soya Flour Mannitol medium (SFM or MS) consisting of 2% soy flour, 2% mannitol, but with glucose instead of mannitol, and 1% peptone similar to the NBP medium. Any other growth medium may be substituted.

The method is as follows:

1. Using a petri dish, inoculate the bacteria sample onto a liquid medium in a petri dish (Note: standing liquid, not agar).
2. The bacteria will form a pellicle (biofilm) at the liquid surface. The author notes that the wrinkled pellicle will look like Japanese *chirimen.* (Note: it's unclear whether the author is referring to the dish made of whitebait, also known as *shirasu*, or the texture of a fabric with the same name of *chirimen*)
3. Discard the liquid below the pellicle by tilting the dish, leaving only the pellicle behind
4. Add silica sand to the pellicle in the petri dish and grind with a pestle.
5. Allow the sand-pellicle mixture to stand at 40°C for 40h. Sporulation occurs during this time.
6. Wash the cells with sterilized water and store as a cell suspension in a cool dark place.

The pellicle formed at the liquid surface helps separate the cells from the liquid culture, so centrifugation is not required. Neither agar work nor shaking equipment is required in this method. At larger scales, culture flasks can be used instead of petri dishes. The silica sand is primarily silicon dioxide (SiO_2), but contains trace amounts of Manganese, which supports sporulation without additional Manganese salts.

This method seems like best practice for commercial spore producers, since it scales well. It is also known that some spore producers sell spores in solid form, mixed with sand. In this case, one can speculate that vacuum desiccation is completed after step 5 in lieu of step 6.

However, it's unclear which of the two methods presented in this section, if any, are used by modern spore producers. This information is not available in the public domain.

Testing Spore Solutions

Spore solutions can easily be tested by heating to 80°C for 30 minutes to kill vegetative cells, diluting, and plating the spore solution to count CFUs.

Spores can also be observed directly with phase contrast microscopy, since there is a large optical difference between vegetative cells and spores.[113] This requires a capable microscope and **counting chamber**, such as the Petroff-Hausser Counting Chamber.[115] This spore count can be used to calibrate the recommended dosage of spore solution for a given quantity of soybeans.

6.4 Genetic Modification Methods

To develop new strains of *B. subtilis* for use in natto, researchers are usually looking to delete, add, or modify certain genes within the *B. subtilis* genome. This section contains an overview of the common techniques used to modify *B. subtilis.* Note that advanced modification techniques, such as CRISPR-Cas9, continues to be developed as the technology advances,[116,117] but as of writing, its use has not been documented on a commercial natto strain.

Modification by Mutation

The crudest methods for genetic modification use random mutation. By subjecting bacteria to sub-lethal doses of radiation and chemicals, mutations can be introduced to the bacterial genome, often leading to the random deactivation of certain genes. Common methods include radiation by ultraviolet (UV) rays, X-rays, gamma rays, and chemical methods using nitro-N-nitrosoguanidine (NTG), ethyl methane sulfonate (EMS), acridine orange (AO), and other chemicals.[113] Mutated bacteria can then be screened for desirable traits.

Modification by Natural Competence and Transformation

The natural competence of *B. subtilis* can be leveraged for genetic modifications. Recall that when *B. subtilis* reaches a certain cell density, a small fraction of the community can enter a state of competence, when they stop reproducing and become able to uptake DNA from their environment. If external DNA is introduced to a competent bacterium, *B. subtilis* can introduce sequences into its own chromosomes through homologous recombination, or through plasmids.[113]

Transduction by Bacteriophages

Genes can be transferred from one cell to another using bacteriophages as an intermediary. When a bacteriophage infects a cell and multiplies, the new phage particle, in rare instances, will incorporate a fragment of the host DNA instead of the phage DNA into its capsule. When this bacteriophage infects another bacterium, it will inject this fragment of DNA, sometimes leading to homologous recombination with the infected bacterium.[118]

7. Case Studies in Natto Strain Development

Researchers have genetically modified commercial and wild type strains of *B. subtilis* to produce novel natto products. These include:

- Low odour natto, for those that dislike the smell.
- Non-sticky natto, for those that don't like the slimy texture.
- Soft natto, for the elderly and those with dysphagia
- High vitamin K2 natto

In this section, we'll examine the exact methods and mechanisms used to develop some of these strains.

7.1 Low Odour Natto *'Kin no Tsubu'* or *'Nioi wa natto'*

Around the year 2000, researchers at *Mizkan* already knew that natto had acceptance issues, even within Japan. Surveys done during that time indicated that most people were eating natto for health reasons, and they had little to no brand preference. Consumers would typically purchase the cheapest package of natto, leading to fierce price competition and a lack of differentiation between brands.

Furthermore, it was well known that natto was more popular in the eastern regions Kanto and Tohoku, but less popular in western Japan, with most people surveyed disliking the odour.

Researchers at Mizkan set out to develop a strain of natto that would have a weaker odour but still confer the health benefits of natto. They speculated that a low odour natto would be easily marketable and set them apart from the competition. Mizkan happened to have an advantage here: the company makes multiple fermented products, including rice vinegar, and had some recent experience developing strains of acetic acid bacteria for vinegar production.[119]

Major odour compounds found in the odour of natto include acetoin, diacetyl, branched chain fatty acids (BCFAs), pyrazines, and ammonia, along with many other odour compounds. Of these, ammonia (which is pungent and offensive) and BCFAs (which has notes of rancidity, cheese, and sweat) are generally perceived as offensive. However, properly produced natto that has been allowed to mature should contain very little residual ammonia, which dissipates during the refrigeration process due to its volatility. Thus, the researchers focused on the removal of BCFAs.

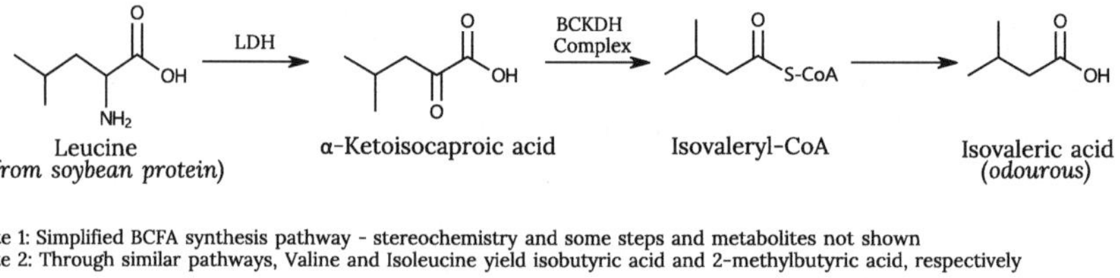

Figure IV.15 BCFA synthesis pathway in B. subtilis.[119]

Isobutyric acid, isovaleric acid, and 2-methylbutyric acid were the key BCFAs targeted for removal. These BCFAs are synthesized by *B. subtilis* from the metabolism of the amino acids valine, leucine, and isoleucine. It was guessed that the first step of this process is facilitated by leucine dehydrogenase (LDH) followed by an aminotransferase (AT), although at the time, the exact mechanism and pathway wasn't fully known.

Theoretically, deactivating either LDH or the AT enzymes should halt BCFA synthesis. To test this, the researchers started with a lab strain of *B. subtilis* designated r22, and introduced a shuttle vector called plasmid pHSG399, a 2227 bp plasmid that inserted itself into the middle of the genes for LDH and AT, respectively, rendering them useless. The LDH deactivated mutant was designated yqiT1, and the AT deactivated mutant was designated ywaA1, after the names of their respective genes. They found that yqiT1, the LDH mutant, generated less than one-fifth the amount of BCFAs when subsequently grown in a liquid culture.

Thus, disrupting LDH appeared to be the more promising approach. The yqiT1 strain was not usable for natto production, however–it was derived from r22 which is a laboratory strain of *B. subtilis* specifically made for genetic manipulation studies, and not for the production of natto.

The next approach was to deactivate the LDH gene from a natto producing strain of *B. subtilis* designated O-2. To delete the LDH encoding gene, the researchers opted to use a combination of homologous recombination and transfection using bacteriophages. In short, modified bacteriophages were used to change a portion of the LDH gene in *Bacillus subtilis* (O-2), inactivating the gene. They named the resultant strain B-2, and it yielded only 0.7mg BCFAs / 100g natto, compared to 70.7mg / 100g natto in the parent O-2 strain.

Table IV.4 Summary of the results from Mizkan's low-odour natto experiments[119]

Strain	Description	Isobutyric acid (mg/L)	Isovaleric acid (mg/L)	Total (mg/L)
Experiments on non-natto producing B. subtilis-on culture media				
r 22	Non-natto strain (parent)	91.8	176.2	268.0
yqiT 1	LDH mutant of r 22	22.7	25.8	48.5
ywaA 1	AT mutant of r 22	125.5	162.7	288.1
Experiments on natto producing B. subtilis-on steamed soybeans				
O-2	Natto strain (parent)	40.0	30.7	70.7
B-2	LDH defective O-2 (transfection)	0.3	0.4	0.7
N46	LDH defective O-2 (mutation)	N/D	N/D	N/D
N64	LDH defective O-2 (mutation)	2.0	2.0	4.0
N103	LDH defective O-2 (mutation)	0.6	0.4	1.0

The original objective was technically achieved at this point, and the research team had developed a natto-producing strain of *B. subtilis* that yielded very low BCFAs. However, the Mizkan team pointed out that commercialization of this B-2 strain might prove to be a challenge. Although the strain is technically sound, the process was still considered genetic engineering, and Genetically Modified Organisms (GMOs) had a bad public perception during this time. The 1999–2001 timeframe included many newsworthy events in the Japanese food industry. This included a drop in sales in 1999 due to concerns over GMO soybeans, and a criticality accident at a nuclear plant near Ibaraki prefecture, close to soybean farms. On the other hand, mad cow disease was headlining news in 2001, and natto sales increased as consumers sought out substitutes for beef.

Regardless, it was clear that the public was highly concerned about food safety and GMOs, and the marketing department at Mizkan was rightly concerned over the use of this newly developed, 'genetically modified' B-2 strain.

Thus, for their final attempt, researchers opted to use the simple but crude chemical mutagenesis, subjecting the original O-2 strain to the chemical N-methyl-N'-nitro-N-nitrosoguanidine (NTG). This is a brute force method, since NTG cannot target individual genes, and researchers had to process 18,896 colonies looking for a random LDH mutant that could still produce high quality natto. Luckily, LDH-deficient strains of *B. subtilis* O-2 were also BCFAs auxotrophs: they required BCFAs for growth. Thus, they could quickly be screened for LDH deficiency using the appropriate culture medium. Three strains (N46, N64, N103) were isolated from this experiment, of which strain N64 had the best natto quality.

Strain N64 was subsequently used to make natto that was branded 'Kin no tsubu (金のつぶ)' and 'Niowanatto (におわなっとう)', and remains one of the most successful cases of natto strain development. It has become one of Mizkan's primary natto offerings alongside its more traditional line "Kume natto (くめ納豆)", and has been hugely successful.

7.2 Non-Sticky Natto '*Mame no ka*'

The Ibaraki Industrial Technology Institute originally sought to create a non-sticky natto to gain acceptance overseas, where natto was notorious for its viscous texture.[120] Compared to the low odour natto developed by Mizkan, this was a relatively simple undertaking. By the time they started in 2013, researchers working with natto were already aware that the commonly used strains of *B. subtilis* var *natto* such as BEST195 were susceptible to a spontaneous loss of sliminess with repeated subculturing.

The mechanism behind this loss of sliminess was elucidated in 2000, when researchers had discovered the IS*4* family insertion sequence, IS*4Bsu*1 (see Section 4.1 and 5.1, this Chapter).[98]

This process happens by chance and doesn't require any external mutation factors or specialized genetic engineering tools. All the researchers had to do was repeatedly subculture the bacteria until it yielded colonies that were not mucoid. In fact, this can be achieved by anyone at home with basic life science tools and techniques, such as those used in home mushroom cultivation. The research paper describing this process demonstrates some of the basic culturing techniques used to isolate strains of *B. subtilis*.

The researchers started with a Miyagino strain, and plated it on a soybean powder agar plate (soybean flour 0.5g, sodium chloride 0.2g, glucose 0.1g, agar 1.5g, 100mL water) and cultured at 37°C for 48h. This was repeated five times. Colonies were then transferred to Luria Broth (LB) agar plates to isolate pure cultures, at 37°C for 24h. Any colonies that appeared different (non-mucoid) were subsequently isolated and grown on Shaeffer agar plates (Difco Nutrient broth 1.2g, magnesium sulphate heptahydrate 0.025g, potassium chloride 0.1g, iron (II) sulfate heptahydrate 0.0278g, chloride tetrahydrate 0.198g, agar 1.5g, water 100mL), which is used to generate spores. These were cultured at 37°C for 48h to obtain sporulated colonies, which were suspended in water to create a starter culture.

Natto tests proceeded based on the methods described earlier, with soaking at 20°C for 16h, draining, cooking at 180kPa for 30mins, and inoculating at 10^3 spores / gram cooked soybeans. This was followed by fermentation at 39°C, 90% RH for 18h, followed by cooling at 20°C, 50% RH for 2h, then curing at 5°C for 24h.

The experiment yielded around 100–200 strains, all of which were tested on steamed soybeans. The resulting natto samples were subsequently tested with several quantitative and qualitative tests (see Chapter III.3). The researchers named their strain 'IBARAKI ℓst–1', and products using this strain were labelled 'Mamenoka (豆乃香)'.

Overall, the new natto product scored similarly to the Miyagino strain and was not significantly different in any of its nutritional or flavour characteristics. The only major differences were the amount of viscous substance produced, its viscosity when stirred, and the perceived stickiness.

Despite their success in the laboratory, by 2017 it was proving difficult to sell Mamenoka in overseas markets, which lack natto producing infrastructure and the culture of eating natto on rice in the morning. Thus, some companies have decided to focus on products derived from natto, including dried (deep fried) natto, dippings, dressings, pastes, and incorporating it into baked goods. Development of these products was also aided by the fact that non-sticky natto is easier to work with for subsequent processing steps.[120]

Table IV.5 Quantitative and qualitative evaluation of low stickiness natto

Parameter	IBARAKI ℓst–1	Miyagino (Parent)
Quantitative Parameters		
Viscous substance (mg/g)	11.5	15.3
Viscosity (RVU)	201	582
Hardness (grams-force)	85.2	88.9
Colour (L*/a*/b*) in CIELAB colour space	55.14 / 5.49 / 11.90	58.97 / 4.55 / 10.32
Vitamin K2 as Menaquinone-7 (ug/100g)	1101	1027
Qualitative Parameters (Natto questionnaire, averages n=38)		
Coverage of Bacteria	2.70	3
Bacteriolytic condition	3.47	3
Cacking and crushing of beans	3.47	3
Colour of beans	3.23	3
Smell	3.23	3
Hardness	3.27	3
Taste	3.13	3
Stickiness	1.47	3
General Score	3.05	3
Qualitative Parameters (Attribute by preference / paired preference, n = 38)		
Which has a better taste? (neither: 11)	13	14
Which has a better smell? (neither: 18)	11	9
Which has stronger stickiness? (neither: 0)	0	38
Which is easier to swallow? (neither: 8)	24	6
Overall, which is better? (neither: 8)	14	16

7.3 Soft Natto

Researchers at the Kyoritsu Women's University originally sought the develop a soft natto to aid individuals who have difficulty chewing and swallowing (a condition known as dysphagia), which can occur in the elderly. This product development did not require any genetic modification techniques, and instead the researchers isolated a novel strain of *B. subtilis* designated KFP843, which was isolated from *sui-douchi*, a natto-like dish from Yunnan Province, China.

Sample natto products were produced from this bacterium using the same techniques described previously, and it was found that the hardness of the resultant natto was about 40% of commercially available natto, such that it could be easily broken apart when stirred with chopsticks.[121]

Subsequent work identified another strain of *B. subtilis*, designated KFP897, that also produced soft natto. However, this natto lacked the characteristic stickiness of natto. To compensate for this, the researchers fermented a co-culture of *B. subtilis* KFP897 with *Micrococcus luteus* IAM1056, which is isolated from miso and creates a sticky texture.

It's unclear whether this product has had commercial success in Japan. However, this study demonstrates the value in using different strains of *B. subtilis*, especially those used to ferment natto-like products overseas, which will be explored in Chapter VII. It also demonstrates the potential of co-culturing entirely different species of bacterium on a sample of natto. These two variables could open an entirely new and novel range of natto products with differing tastes and textures.

Summary

To summarize, *Bacillus subtilis* is a well-studied bacterium, but not all strains of *B. subtilis* produce good natto. So called 'natto-producing strains' can be characterized by the production of γ-PGA and the 'aroma of natto', but clearly γ-PGA is not always desirable, and the 'aroma of natto' is subjective at best. Many studies have elucidated the inner workings of *B. subtilis*, and researchers have been able to breed strains of *B. subtilis* that produce natto with specific qualities. Novel wild strains of *B. subtilis*, and even strains of entirely different species, should continue to be tested.

Further Reading

Cellular and molecular biology is a complex subject, and the contents of this chapter only skim the surface. Some further reading is provided below.

For a basic understanding of biology and bacteriology, I encourage the reader to start with any secondary school level textbook such as:

- **Microbiology: An Evolving Science, 5e** by Joan L Slonczewski, John W Foster, and Erik R. Zinser.[122]

From there, an advanced understanding of molecular biology in bacteria can be found in:

- **Snyder and Champness Molecular Genetics of Bacteria, 5e** by Joseph E. Peters and Tina Henkin[123]

For heavy details specifically on *B. subtilis*, readers should stay up to date with academic literature and *B. subtilis* textbooks such as:

- **Bacillus subtilis and Its Closest Relatives: From Genes to Cells**, edited by Abraham L. Sonenshein, James A. Hoch, and Richard Losick.[124]–This book has a greater focus on metabolic pathways.
- **Bacillus: Cellular and Molecular Biology, 3e** edited by Peter L. Graumann.[125]–This book has a greater focus on molecular machinery.

Modern Japanese language texts on Natto science include:

- 納豆の科学 [**Advanced Science on Natto**] by Kiuchi Kan et al.[126]
- 納豆の研究法 [**Investigative Approach to Natto**] edited by Kiuchi Kan, Toshiro Nagai, and Keitaro Kimura.[127]

V. The Soybean and its Biochemistry in Natto

From a culinary and agricultural standpoint, soybeans (*Glycine max* (L.) Merr.) are a double-edged sword. On one hand, they're an agricultural powerhouse, yielding more protein per acre on productive farmland than any other crop, and can harbour nitrogen-fixing bacteria to boot. On the other hand, they're a culinary challenge: soybeans can take an incredibly long time to cook and are unappetizing at best. Its culinary history is full of techniques used to transform it into something palatable.

The taxonomical species *Glycine max* is usually written down as *Glycine max* (L.) Merr., where the latter two abbreviations refer to the Swedish botanist Carl Linnaeus (responsible for the genus name, *Glycine*) and American botanist Elmer Drew Merrill (responsible for the species name, *max*), respectively. Following millennia of domestication and breeding, *Glycine max* is incredibly different than its wild predecessor (*Glycine soya* Sieb. & Zucc.), which produces tiny black seeds weighting <3g/100seeds, compared to around 9g/100 seeds for natto grade soybeans and around 20g/100 seeds for food grade applications.

Learning about the soybean allows us to understand the biochemical transformations during the natto production process. It allows us to troubleshoot bad natto, develop new natto products, work with different cultivars of soybeans, different beans entirely, and how to select beans suitable for natto.

The same knowledge also helps our understanding of the olfactory, compositional, and physiological aspects of natto, which can be used to improve natto's taste, nutrition, and functionality.

In this chapter, we'll take a closer look at the soybean and the how the individual components of the soybean seed, including its carbohydrates, proteins, fats, and vitamins change during the natto making process.

1. The Soybean (*Glycine max*)

Soybeans are a member of the family *Fabaceae*, which encompasses all plants commonly referred to as legumes, peas, or beans. Native to East Asia, the soybean has a huge agronomical importance, owing to its high yield and protein content.[128] In East Asia, its literal name usually translates to 'yellow bean' or 'big bean', where as the English name 'soybean' is likely a corruption of the Japanese word *shoyu* (soy sauce).

Table V.1 Proximate composition of common legumes, per 100g dry weight

Common name	Scientific name	Proteins	Lipids	Carbohydrates, total	Dietary fibre	Ash
Peanut	*Arachis hypogaea*	27.0	50.5	20.0	7.9	2.4
Adzuki bean	*Vigna angularis*	24.2	2.3	69.5	28.9	4.0
Kidney bean	*Phaseolus vulgaris*	26.1	3.0	66.6	23.1	4.4
Pea	*Pisum sativum*	25.1	2.7	69.7	20.1	2.5
Cowpea	*Vigna unguiculata*	28.3	2.4	65.1	21.8	4.3
Broad bean	*Vicia faba*	30.0	2.3	64.5	10.7	3.2
Soybean (Japanese)	*Glycine max*	38.6	22.5	33.7	24.5	5.4
Rice bean	*Vigna umbellata*	23.6	1.8	70.2	25.0	4.4
Chickpea	*Cicer arietinum*	22.3	5.8	68.6	18.2	3.2
Scarlet runner bean	*Phaseolus coccineus*	20.3	2.0	72.3	31.6	5.3
Lima bean	*Phaseolus lunatus*	24.8	2.0	68.9	22.2	4.3
Mung bean	*Vigna radiata*	28.1	1.7	66.3	16.4	3.9
Lentil	*Lens culinaris*	26.4	1.7	69.0	19.0	3.1

Note 1: values for raw beans, from Standard Tables of Food Composition in Japan 2020 (8th revision)[129], corrected for 100g dry weight (no moisture). Total carbohydrates include dietary fibre. Dietary fibre includes oligosaccharides, and the raffinose family of sugars.
Note 2: the USDA also publishes reference values for food composition, but the Japanese Standard Tables will be used since it also includes accurate data on natto. The USDA record for natto is based on 'similar' foods–likely cooked soybeans–rather than measuring natto itself.

The soybean's composition is quite different from members of the bean family, such as common beans (*Phasoleus spp.*), lentils (*Lens spp.*), or chickpeas (*Cicer spp.*) which contain comparatively high amounts of starch, and a very small amount of lipids.[130] Soybeans, on the other hand, contain large amounts of proteins and lipids, with a more balanced amino acid profile for human nutrition. The large amount of lipids puts soybeans into the commodity category of an oilseed, although some have argued that it should be classified uniquely as a "protein-seed" instead.[131]

1.1 Seed Morphology

Seed Structure

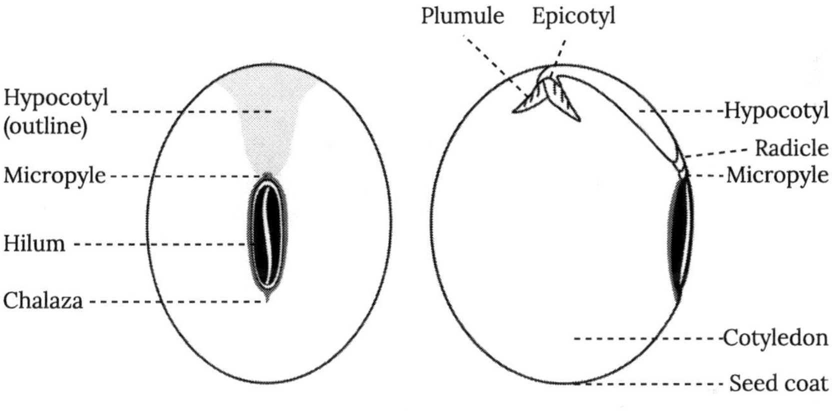

Figure V.1 Soybean seed structure.
Left: view facing hilum. Right: cross section if split along the cotyledon halves.

If we take a closer look at a mature soybean seed, we can see that it's made primarily of two basic parts: the seed coat, and the embryo. This is different from agricultural crops of the grass family *Poaceae* such as rice and corn, in which we would also expect to see a large starchy energy reserve called the endosperm. Soybeans don't have a significant endosperm, and within the seed coat, the remainder of the soybean forms the embryo.

The function of the seed coat is to protect the seed from fungi and bacteria before and after planting, while also allowing the soybean to soak up moisture. The seed coat is marked on one side by the hilum, which is a linear or ovular scar on one side of the seed marking the point of attachment to the seed pod.[128] Located on opposite sides of the hilum are a small hole called the micropyle, which is the original point of pollen entry into the soybean ovule, and the chalaza.

The largest component of the embryo is the cotyledon, which is the primary reserve of food and nutrients for a germinating soybean plant. Once germinated, the cotyledon becomes the first 'seed leaves' of the plant, after which true leaves will grow. The remainder of the embryo is located on one side of the hilum, and consists of the radicle, hypocotyl, and epicotyl.[128] Collectively, these components are also called the hypocotyl axis, or germ, and make up only a small portion of the seed.

Seed Cellular Structure

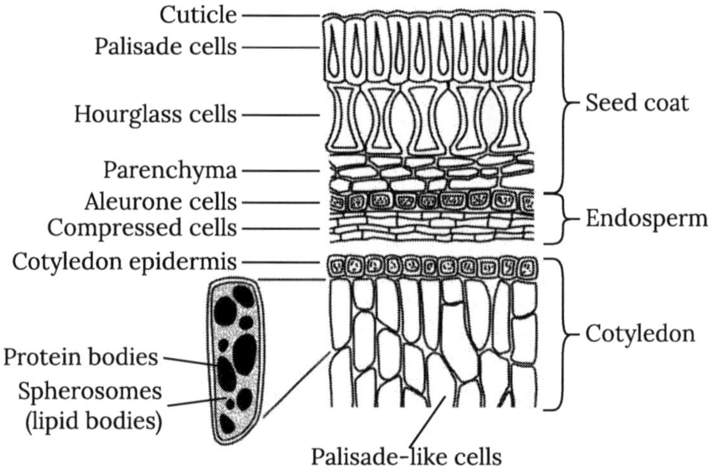

Figure V.2 Soybean cross section at cotyledon

As a raw material, we are mainly concerned with the seed coat and cotyledons, which collectively make up 98% of the soybean's mass. If we zoom into a cross section of the seed coat located at the cotyledon, we can see that it's made of multiple layers of specialized cells.

Starting from the surface of the soybean, the seed coat is comprised of the protective cuticle, then palisade cells, hourglass cells, then layers of parenchyma cells.

Underneath the seedcoat, there is a layer of aleurone cells and compressed endosperm tissue, which make up the 'endosperm' of the soybean.

Finally, the cotyledon consists of an epidermis, and the interior is filled with elongated palisade-like cells, which contain most of the valuable oil and protein. These are contained in cellular structures known as spherosomes (or oil bodies) and protein bodies, respectively.[128,132] Despite being about 22% lipids by dry weight, soybeans don't feel 'greasy' when consumed because the spherosomes remain intact. These spherosomes are ruptured during soybean oil extraction.

1.2 Seed Growth

Germination

If you place the soybean into a moist environment, it will absorb water, swelling to double in size. During germination and growth, the radicle becomes the primary root of the plant. The hypocotyl lifts the cotyledons above the soil, and forms the portion of the stem below the cotyledons. The epicotyl forms the main stem above the cotyledon, eventually yielding the first true leaves.

Agronomists usually divide the soybean's growth into two main phases: vegetative growth, which involves leaves and nodes, and reproductive growth, which includes the formation of flowers, pods, and seeds.

Vegetative Growth (Vn)

During vegetative growth, the cotyledons emerge (VE, emergence stage), followed by unifoliate true leaves (VC, unifoliate stage), after which sets of trifoliate leaves grow (stage Vn, where n=sets of trifoliate leaves). When planted optimally, the plant will develop 5–7 trifoliates (stages V5–V7) before flowers emerge.[133]

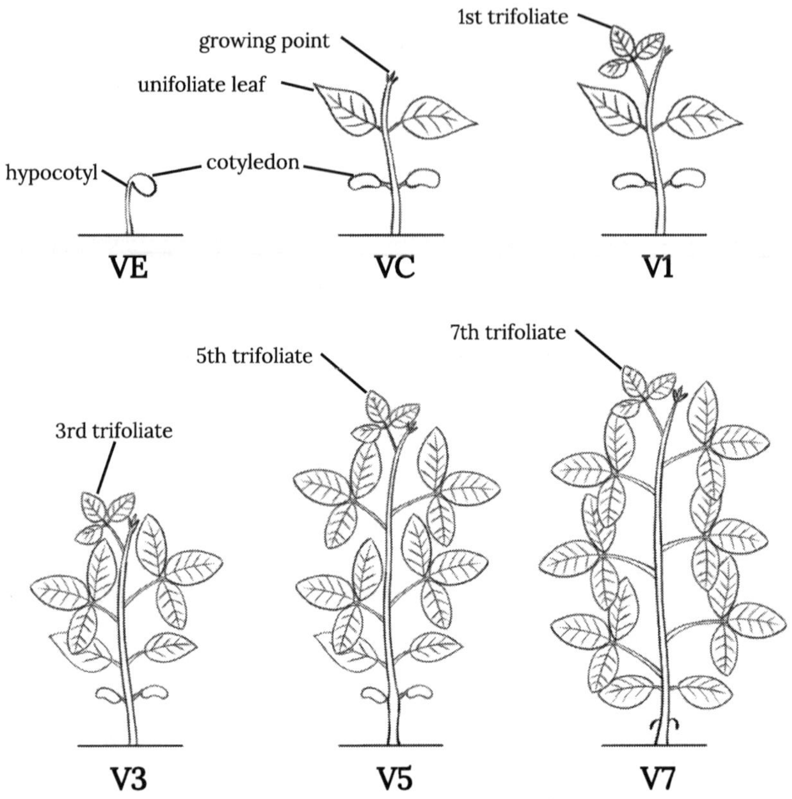

Figure V.3 Soybean vegetative growth stages

Reproductive Growth (Rn)

The flowering process is triggered by day length and temperature changes, but varies with cultivar. The reproductive stages (R1–R8) are distinguished by the state of flowers and bean pods. The first stage, R1, occurs when one flower is present on the plant. Several hair-covered bean pods will develop at the plant's nodes, eventually elongating and developing seeds with a high sugar and moisture content. At R6, the seeds reach their maximum size and can be eaten fresh as *edamame*. From there, the plant begins to wither and dry out as the seeds mature. At R8, the seeds are fully mature and ready for harvest.[133]

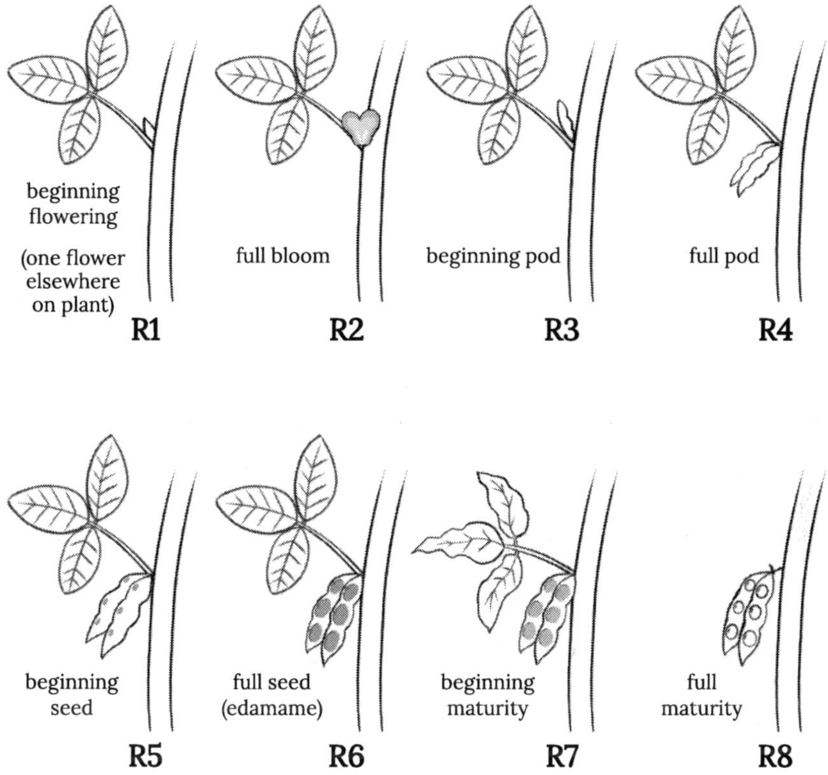

Figure V.4 Soybean reproductive growth stages

Nitrogen Fixation

One of the most important qualities of all beans, including soybeans, is their ability to form root nodules by a symbiotic association with nitrogen-fixing bacteria, such as *Rhizobia spp.* These bacteria take atmospheric nitrogen gas, N_2, and fix it into ammonium-N (NH_4^+/NH_3), which can be directly utilized by the plant.[134]

These nodules typically appear 9 days after planting, and start nitrogen fixation at around 14 days. This nitrogen-fixation system provides as much as 90% of the fixed nitrogen to the soybean plant, meeting its full nitrogen requirements and resulting in its high protein yield per acre. The ability to fix nitrogen, among other factors, makes soybeans the preferred candidate crop to corn and wheat in crop rotation systems.[134]

1.3 Physical Characteristics

Key physical characteristics of the soybean are its size, colour, and water absorption. Natto producers characterize soybean size by their diameter or ability to pass through a certain screen size, while farmers characterize soybean size by the weight of 100 individual beans. Refer to Chapter II.1.1 for a full description of soybean size classifications.

In practice, natto soybeans and sprouting soybeans are the two major end-use applications that specifically require small soybeans. All other applications tend to use larger beans, since these will yield more beans per acre of land.

Mature soybeans seeds can be yellow, black, brown, green, or a bicolour of these, which is based on the presence or absence of pigments, flavonoids and anthocyanins in the seed coat, and is determined by several genetic markers.[135] The colour of the hilum can range from yellow, gray, brown, or black, and is a useful trait in identifying the cultivar of soybean.[136] Traditionally, natto producers have preferred yellow soybeans with light yellow hila, but producers have been experimenting with more unique varieties. Anecdotally, black soybeans tend to have a more subdued flavour and odour when made into natto.

2. Soybean Composition and Natto Fermentation

2.1 Proximate Composition

Table V.2 Composition of each component of the soybean seed.[132]

Component	Weight	Protein	Lipids	Ash	Carb.
Whole bean	100.0	40.3	21.0	4.9	33.9
Cotyledon	90.3	42.8	22.8	5.0	29.4
Seed Coat	7.3	8.8	1.0	4.3	85.9
Hypocotyl Axis	2.4	40.8	11.4	4.4	43.4

The cotyledons comprise most of the soybean's mass (90.3%), followed by the seed coat (7.3%) and hypocotyl axis (2.4%). Both the cotyledons and hypocotyl axis have significant amounts of proteins, lipids, and carbohydrates, whereas the seed coat is primarily carbohydrates in the form of fibre.[131,132]

The whole bean composition and size of soybeans varies by cultivar and growing conditions, and the choice of cultivar is also affected by the end use application. Different soybean compositions and sizes are desirable for tofu, miso, natto, oilseeds, and animal feed, among others.[128]

2.2 Carbohydrates

Carbohydrates are the least economically valuable component of the soybean, but are one of the most important parameters when selecting soybeans for natto.

Soybeans carbohydrates are classified into soluble and non-soluble carbohydrates. Natto producers want soybeans with a large amount of soluble carbohydrates to feed the growing bacterium. *Bacillus subtilis* is generally poor at breaking down starch and insoluble fibre in soybeans, so it relies on soluble sugars as its primary energy source for growth and metabolism.

Soluble Carbohydrates

The main soluble carbohydrates in soybeans are sucrose, raffinose, stachyose, and verbascose. Sucrose is a disaccharide of glucose and fructose, whereas the latter three sugars are known as Raffinose Family Oligosaccharides (RFOs) and are composed of sucrose bonded to one, two, or three molecules of galactose through α-1,6-glycosidic bonds.[137]

Humans cannot digest RFOs since we lack the enzyme α-galactosidase needed to break the bonds between galactose molecules, and between galactose and sucrose. RFOs normally pass into our large intestine, where gut bacteria digest it as dietary fibre, causing flatulence.[130,138] The remaining carbohydrates are monosaccharides, mainly glucose and arabinose, which occur in trace amounts in soybeans. Some carbohydrates are also bound to proteins and lipids, forming glycoproteins and glycolipids, respectively.

In natto, *B. subtilis* has the ability to break down all of these oligosaccharides into simple sugars, metabolizing them for energy or converting them into different substances.[139]

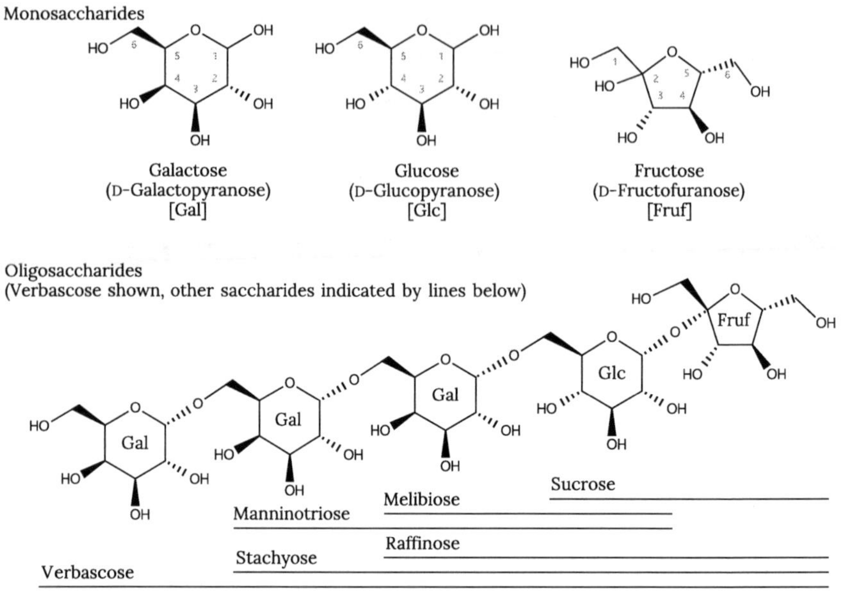

Figure V.5 Selected mono- and oligosaccharides found in soybeans and natto.

Digestion of RFOs by α-galactosidase removes the galactose from one end of the oligosaccharide, eventually leaving behind sucrose, which is broken down into glucose and fructose. For this reason, the concentration of glucose and fructose temporarily increases around the 8h mark of fermentation, before decreasing again as the sugars are consumed by growing *B. subtilis.*

Table V.3 Carbohydrate content of beans during natto making process (dry weight %).[140]

Carbohydrate	Raw Soybean	Soaked Bean	Steamed Bean	Natto
Sucrose	7.05	6.01	5.74	tr
Raffinose	1.15	0.97	0.85	tr
Stachyose	6.34	5.58	5.09	0.41
Melibiose	tr	tr	tr	0.16
Manninotriose	tr	0.06	0.12	1.18
Glucose	nd	tr	0.23	tr
Fructose	nd	tr	0.14	nd
Myo-inositol	0.07	0.07	tr	tr
Starch	0.98	0.99	0.62	0.64

tr = trace, <0.05%; nd = not detected

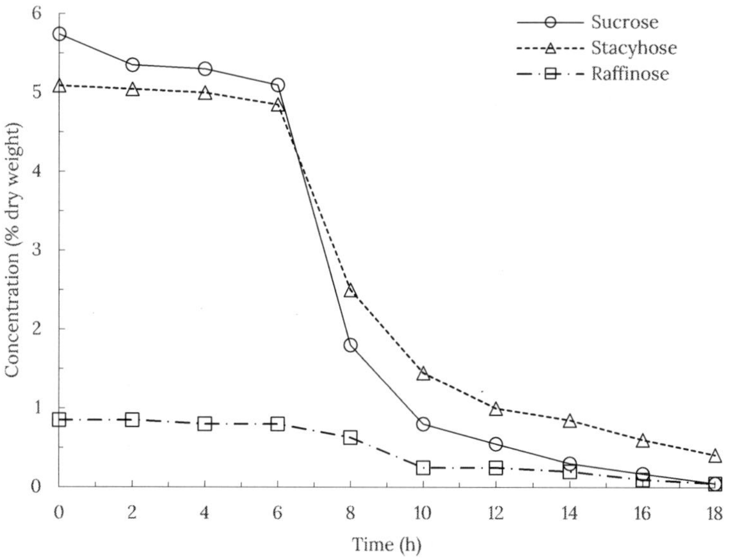

Figure V.6 Changes in carbohydrates during natto fermentation (RFOs and sucrose).[140]

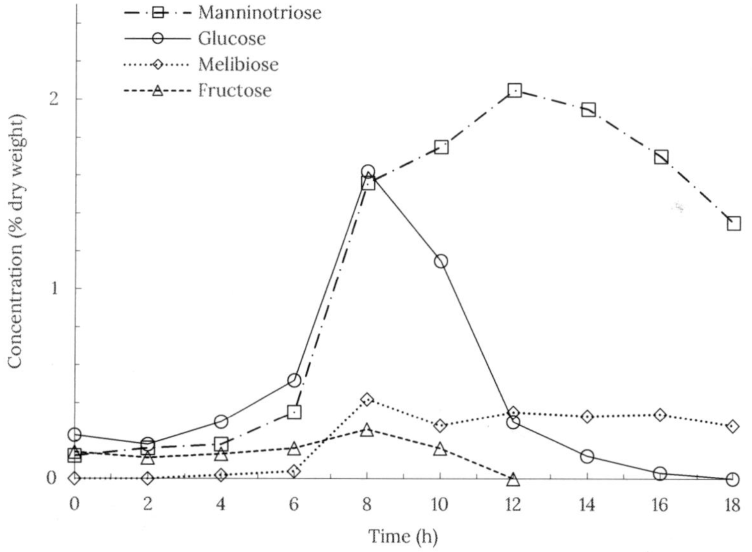

Figure V.7 Changes in carbohydrates during natto fermentation (degradation products).[140]

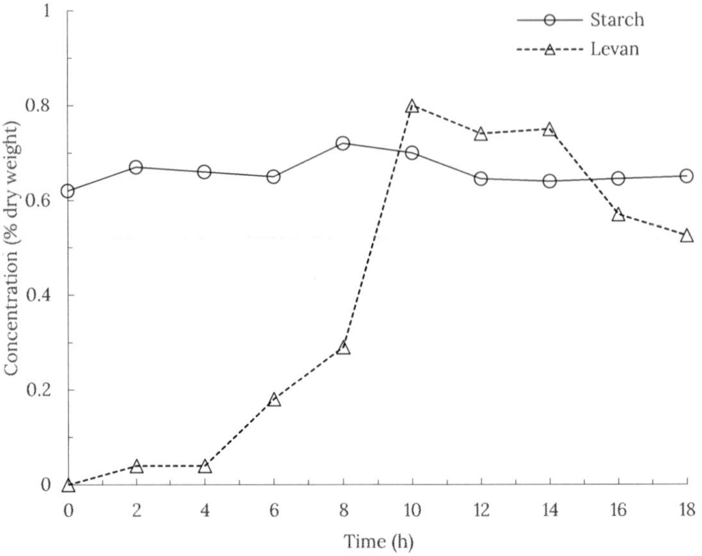

Figure V.8 Changes in carbohydrates during natto fermentation (polymers).[140]

Digestion of these sugars by sucrase removes the fructose from the opposite end of the oligosaccharide, leaving behind intermediate sugars manninotriose and melibiose, which are digested more slowly.

Excess sucrose is used by *B. subtilis* to form levan, a fructose polymer that is one of the main constituents of the bacterial biofilm. The formation of levan liberates glucose.

Overall, you can observe a rapid decline in sucrose, raffinose and stachyose around the 8-h mark as exponential growth of *B. subtilis* progresses, coinciding with a temporary spike in the degradation products and levan. All sugars subsequently decline as *B. subtilis* uses them for growth.[140]

Insoluble Carbohydrates

The main insoluble carbohydrates in soybeans are hemicellulose, cellulose, and pectin–all of which are forms of insoluble fibre. Soybeans also contain starch, which is digestible by humans but only makes up a small fraction of soybean carbohydrates.

Cellulose, hemicellulose and pectin are primarily structural carbohydrates, making up the seed coat and the cell walls within the soybean. Starch is a storage polysaccharide found in cotyledon cells, providing an energy source for the growing soybean.[131,137]

In natto, *B. subtilis* cannot effectively break down any of the insoluble carbohydrates, and they remain in the final product.[140]

2.3 Proteins

Proteins are one of the most valuable parts of the soybean, although the exact quantity and composition of protein is less important for natto-making than carbohydrates. These proteins come in all shapes and sizes, ranging from single peptides to complexes containing multiple subunits.

Proteins are comprised of amino acids, which polymerize and fold into many shapes and sizes with different functions and properties. As complex polymers, proteins are more difficult to classify compared to starches and lipids.

Classification Based on Function

The easiest way to think about proteins in the context of soybean seeds is by their function. Soybean proteins can be grouped by function into metabolic proteins or storage proteins.

Metabolic proteins such as enzymes are needed to support the reactions within the plant's physiology. The main metabolic proteins found in soybeans are Hemagglutinin (Lectin), Trypsin inhibitors, Lipoxygenases, α-amylase, and Cytochrome-C. Of course, *Glycine max* contains an entire genome coding for many more proteins, but these are insignificant when we look at the composition of the mature soybean seed.

Storage proteins have no function other than to provide a source of food during the plant's early development. The main storage proteins of the soybean are β-conglycinin and glycinin. By mass, the storage proteins make up the bulk of the soybean seed and are found in the protein bodies of the palisade-like cells inside the cotyledon.

Classification Based on Solubility and Family

Proteins can also be classified based on solubility. If the protein is soluble in water, we classify it as an albumin, and if it's soluble in a dilute salt solution, it's classified as a globulin. Most of the soybean's proteins are globulins.

For legumes, agricultural chemists further distinguish storage globulins into two families of storage proteins: legumin, which has a large molecular weight, and vicilin, which has a smaller molecular weight.

Classification Based on Svedburg Units

If we take the protein, solubilized in a salt solution, then put it in an ultracentrifuge, we can separate the proteins into fractions based on how quickly they settle within the centrifuge, i.e. its sedimentation rate.

The Svedberg unit, S, with units of 10^{-13} seconds, is commonly used in biochemistry as a measure of a particle's sedimentation rate and is calculated with the following.[131]

$S\,[10^{-13}\,s] = (dc/dt)r\omega^2$

Where,

dc/dt is the particle velocity

$r\omega^2$ is the centrifugal acceleration

Thus, a protein of sedimentation rate 2S (2×10^{-13}s) would travel 2 micrometers per second (2×10^{-6} m/s) when subjected to a centrifugal acceleration of 10^7 ms^{-2} (slightly over 10^6 gravities, or 10^6G).

This unit is widely used in biology, and is dependent on the size, shape, and charge of the molecule being measured. Recall that this is the same unit used for measuring the size of ribosomal subunits in Chapter IV.5.1. Svedberg units aren't linear or additive, since they only account for settling time, not the actual molecular weight of the protein (which is measured in kilodaltons, or kDa). Nevertheless, Svedberg units are mentally interpreted as size.

If we dissolve the soybean protein in a 0.5 molar (M) salt solution and centrifuge it, the protein will separate into four fractions (based on their settling velocity in Svedberg units): 2S, 7S, 11S, and 15S, each containing different functional and storage proteins.[131,141]

It's important to remember the designations 2S, 7S, 11S, and 15S always refer to fractions of proteins based on their sedimentation rates, not the proteins themselves. In literature, authors will often refer to proteins by their Svedberg units (e.g. 7S soybean protein) which can lead to confusion.

The Svedberg unit also has limitations since it depends on the salinity of the solution. For example, if the same soybean protein is dissolved in a 0.1M salt solution rather than 0.5M, an additional 9S fraction appears, containing a dimerized form of some of the 7S proteins.[131]

Overview of Storage Proteins

Storage proteins account for 60–80% of the total protein in soybeans, and they are soluble in salt solutions (they are Globulins). The smaller protein is classified as a vicilin, is found in the 7S fraction, and the name of the protein is β-conglycinin. The larger protein is classified as a legumin, is found in the 11S and 15S fractions, and consists of glycinin.[131,141]

β-conglycinin

The β-conglycinin storage protein is a trimer consisting of three subunits that are called α, α', and β subunits. Mathematically, there are ten possible ways these three subunits can be arranged into a trimer, although only six have been found (B1–B6).

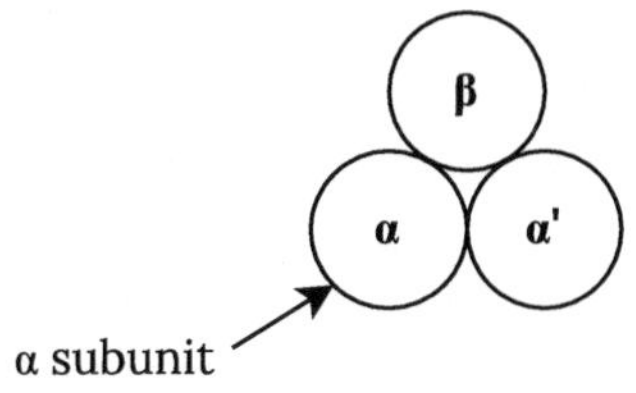

Figure V.9 β-conglycinin (B3 trimer, 7S)

Table V.4 β-conglycinin subunits and trimer configurations

	Subunit		
Subunit Properties	α	α'	β
MW (kDa)	67	63	48
No. Methionine	1	4	0
No. Cysteine	1	1	0
Known Trimers			
B1		1	2
B2	1		2
B3	1	1	1
B4	2		1
B5	2	1	
B6	3		

The α and α' subunits are low in the amino acid methionine, whereas the β subunit is devoid of cysteine and methionine. It was originally thought that α and α' subunits contained no cysteine, but it was later found that roughly half of α and α' subunits are connected by disulfide bonds, either directly to each other or through an intermediate protein called P34. Based on amino acid analysis, the nutritional value of the subunits is α'> α > β in order of descending value of essential amino acids.[131,141]

This trimer form of β-conglycinin can rearrange itself into a hexamer under low ionic strength salt conditions, with two trimers facing each other. This configuration shows up as the 9S fraction in an ultracentrifuge in 0.1M salt solutions.

Furthermore, β-conglycinin is a glycoprotein, and carbohydrates of mannose and glucosamine account for roughly 5% of the total weight.

Glycinin

Glycinin is the larger storage protein, and forms the largest fraction of total soybean proteins. Compared to β-conglycinin, it is richer in essential amino acids and only a small amount is glycosylated.

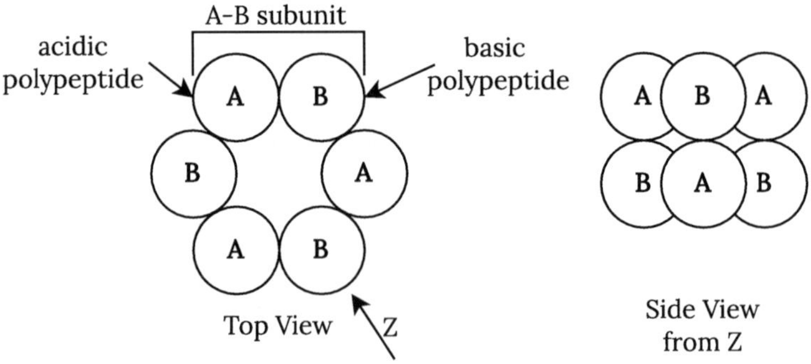

Figure V.10 Glycinin hexamer

Glycinin is a hexamer with a molecular weight of 360 kDa, and each of its six subunits consists of an acidic polypeptide (denoted Ax) and a basic polypeptide (denoted Bx), connected to each other by a single disulfide bond. In total, five A–B subunits have been identified as subunit G1–G5, with G5 being an exception to the rule (it contains two A polypeptides and one B polypeptide). The G1–G3 subunits contain more methionine than G4 and G5, making them more nutritionally desirable.[131,141]

Table V.5 Glycinin subunits and polypeptides

Subunit	Acidic polypeptide	Basic polypeptide	MW (kDa)	No. Methionine
G1	A_{1a}	B_2	58	5–6
G2	A_{1b}	B_{1b}	58	5–6
G3	A_2	B_{1a}	58	7–8
G4	A_3	B_4	62	3
G5	A_5A_4	B_3	69	3

Importance of the Storage Proteins for Natto

Overall, the two storage proteins, β-conglycinin and glycinin, comprise between 65–80% of the soybean seed's protein, and these two proteins have the largest impact on the soybean's texture, nutrition, thermal properties, and other properties when made into food products. Glycinin is more nutritionally important than β-conglycinin, and contains between 3–4 times as much methionine and cysteine than the latter.[131,141]

Overview of Functional Proteins

Functional proteins are separated in the 2S and 7S fractions. Almost all of these proteins have undesirable physiological effects on the human body, but these effects are mitigated by cooking, which denatures the nascent protein.[131,141]

Lectins

Soybean lectins, also known as hemagglutinins or soybean agglutinins, are proteins that bind to carbohydrates. They show up in the 7S fraction, have a molecular weight of 120kDa, with four identical subunits of 30kDa each. They are glycoproteins, and contain around 5% mannose and glucosamine.[131,141]

Because of their ability to bind to carbohydrates, they can bind to the carbohydrates on the surfaces of human red blood cells, causing them to agglutinate, hence the name hemagglutinins. In some legumes such as kidney beans containing the lectin *Phytohemagglutinin*, these lectins can be highly toxic. The effects of soy Lectins are milder, and can cause inflammation, malabsorption of protein, and other effects if consumed raw.

Soybean Lectins are easily denatured by heat in the presence of water, which renders them harmless.

Trypsin Inhibitors

Trypsin inhibitors are proteins that bind to serine proteases, which are enzymes used by both eukaryotes and prokaryotes to break down proteins by cleaving peptide bonds. Because of this effect, trypsin inhibitors are classified as anti-nutritional factors, and can stunt the digestion and absorption of proteins in animals.[131,141]

Like lectins, trypsin inhibitors are denatured by cooking, although they are more resilient than lectins. Higher temperatures speed up the denaturing process of soybean trypsin inhibitors.

Lipoxygenases

Lipoxygenases are iron-containing proteins that catalyze the oxygenation reaction of a double-bond in polyunsaturated fatty acids, yielding a hydroperoxide. With the help of other enzymes such as hydroperoxide lyase or hydroperoxide isomerase, the hydroperoxide can split into many six- and nine-carbon compounds, creating a 'beany' odour that some regard as undesirable.[131,141]

This 'beany' or 'green' odour is a problem in tofu and soymilk production, where grinding raw, soaked soybeans brings lipoxygenases, polyunsaturated fatty acids, and oxygen into close contact with one another prior to heating. In East Asian cultures, this beany aroma can be considered desirable, while many western soymilk producers will mitigate the effect of lipoxygenases by grinding soybeans with hot water.

Degradation of Soybean Protein for Natto

The exact protein composition of soybeans doesn't have a significant effect on the growth of *Bacillus subtilis* in natto. All the major functional proteins, including lectins, trypsin inhibitors, and lipoxygenases, generally have negative effects on nutrition and taste, but these proteins are partially removed during soaking and denatured during the pressure-steaming process. Storage proteins make up the bulk of the soybean's composition, and these are also denatured during the steaming process.

B. subtilis excretes 'feeding proteases' that dissolve into the cooked soybean and digest the denatured protein into smaller peptides and amino acids. *B. subtilis* secretes two major feeding proteases: AprE, a 39kDa serine protease (subtilisin E), and NprE, which is a 56kDa neutral metalloprotease.[142] In the natto-producing strain BEST195, AprN (subtilisin NAT) is secreted instead of AprE. Inactivating the genes for both proteases reduces the protein degrading ability of the bacteria by over 96 percent.[143]

This digestion process can be observed by measuring degradation products like amino nitrogen, ammonia caused by the deamination of amino acids, and protein size through gel electrophoresis. The solubility of proteins will also increase as they are broken down into peptides, and the generation of ammonia increases the pH of the steamed soybeans from slightly acidic (pH 6.5) to slightly basic (pH 7.7).[144,145] These parameters were measured in separate experiments in 1964 and 1984.

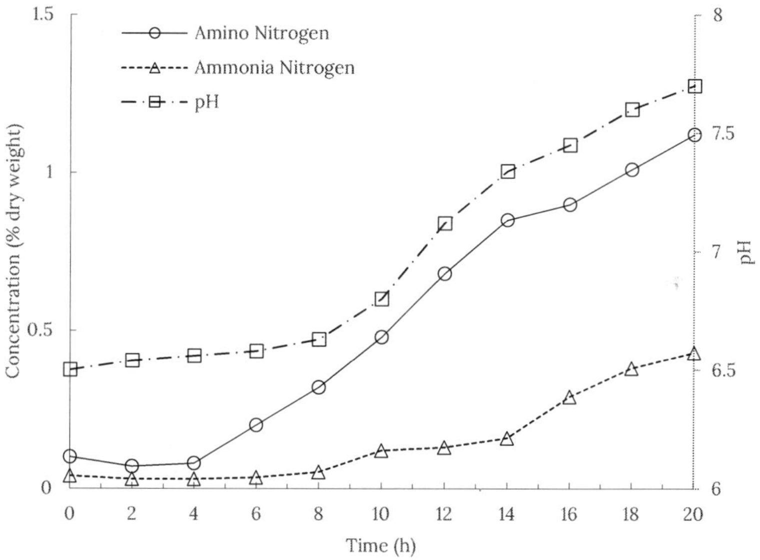

Figure V.11 Changes in amino-N, ammonia-N, and pH during natto fermentation.[145]

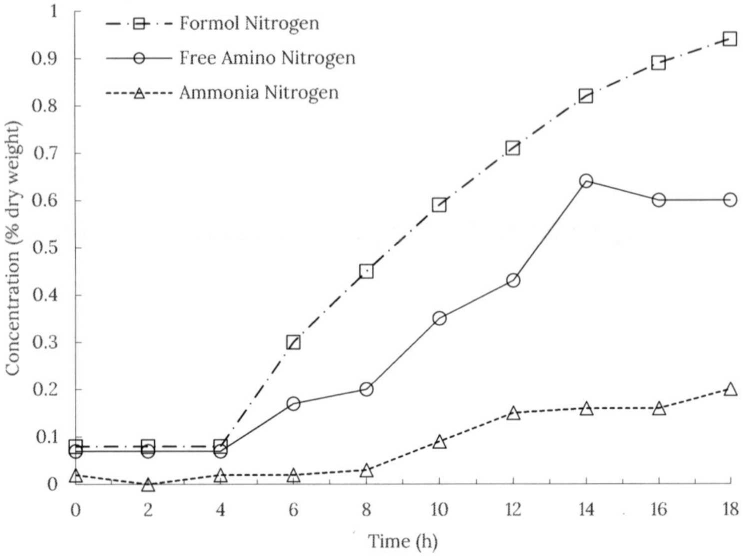

Figure V.12 Changes in ammonia and amino nitrogen during natto fermentation.[144]
Note that Sørensen formol titration is rarely used today.

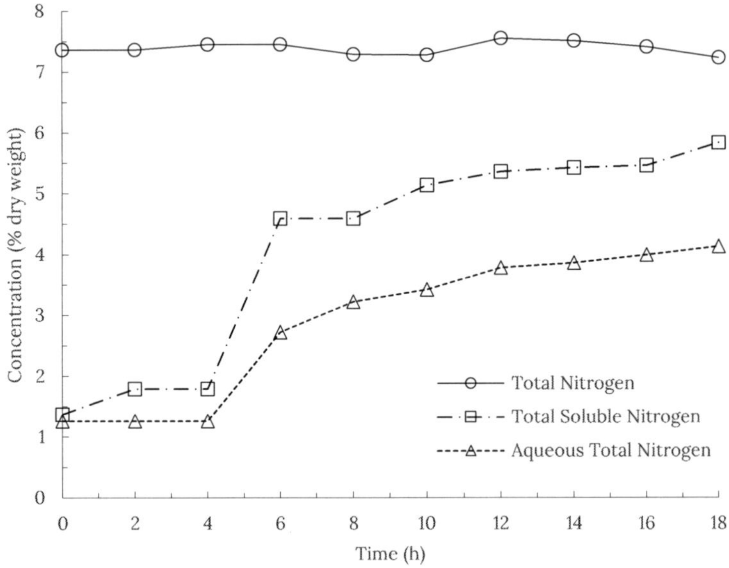

Figure V.13 Changes in nitrogen compounds during natto fermentation.[144]

2.4 Energy and Bacterial Count

Carbohydrates and proteins are the two major energy sources for *B. subtilis* during natto fermentation. As these compounds are broken down for energy, heat will be released proportional to the bacterial population. The evolution of heat rapidly increases at the 8h mark, after the lag phase, and reaches a maximum between 12–24h.[146]

Crushing the soybean into smaller pieces, in the case of *hikiwari natto*, increases the maximum rate of heat production and causes peak heat production to occur earlier in the process. The same is true if the seed coat is physically removed from the steamed soybean. Both of these factors allow bacteria easier access to proteins and carbohydrates within the soybean.[146]

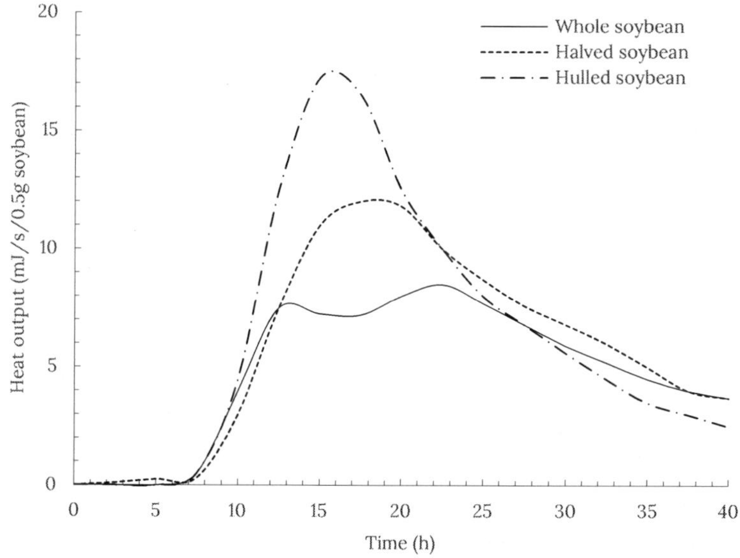

Figure V.14 Heat output of 0.5g natto over extended fermentation period.[146]

The amount of heat emitted is also affected by the temperature of fermentation, which directly influences the bacterium's growth rate. Maximum heat output occurs when the natto is fermented around 40°C. At lower temperatures, growth is much slower so the natto takes longer to reach peak heat output. Fermenting natto as low as 27°C means that peak heat output won't occur until 40h into growth, and the maximum heat output is much lower.[146]

Total bacteria count, measured in colony forming units (CFUs), increases exponentially during throughout fermentation. Natto is typically inoculated at 10^4 CFU/g soybeans. During the first 8h of growth, the bacteria population increases rapidly, increasing tenfold about every two hours. After 8h, this increase tapers off, and the population will increase tenfold about every 5h until the completion of fermentation.[147] Once all of the spores in the inoculum have germinated, only a very small fraction (<1%) of total cells will be spores during the first 8h of growth. The amount of spores versus vegetative cells subsequently increases, and roughly 10–40% of the *B. subtilis* cells will be present as endospores at the end of fermentation.[140] The total cell and spore count varies considerably between experiments, and final cell counts range from 10^{10}–10^{15} CFU/g.

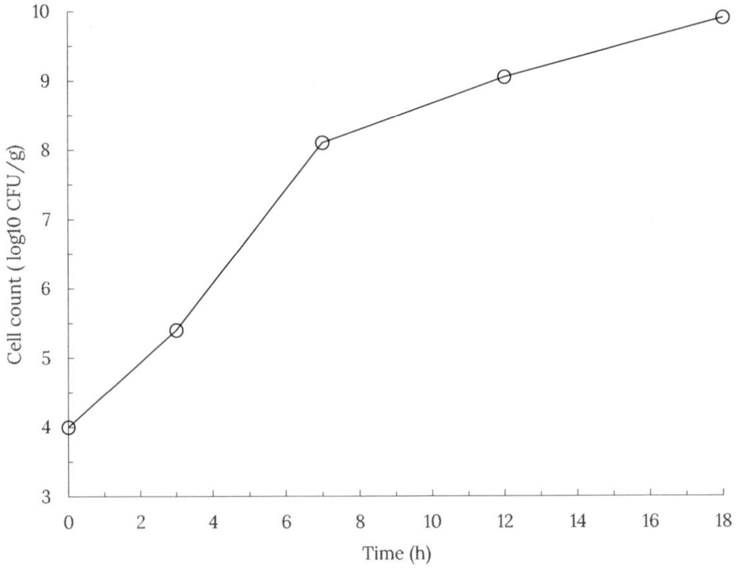

Figure V.15 Total cell count during fermentation.[147]

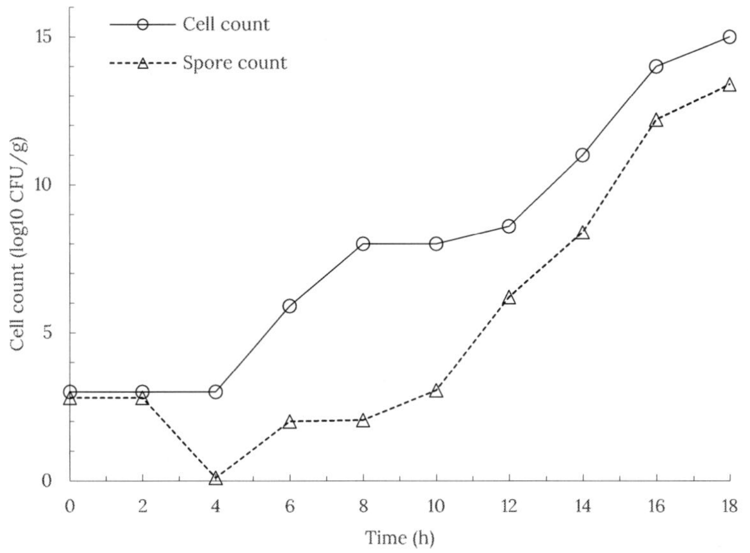

Figure V.16 Total cell count and spore (heat shocked) cell count during fermentation.[140]

2.5 Lipids

The lipids of soybeans are primarily found in oil bodies, also known as lipid bodies, spherosomes, oleosomes, or lipid-containing vesicles, within the palisade-like cells of the cotyledon. Triglycerides comprise 95–97% of the total lipids, followed by small amounts of phospholipids, free fatty acids, and unsaponifiable matter (sterols, tocopherols, and hydrocarbons).[131,148]

Triglycerides and Fatty Acids

Triglycerides consist of a glycerol backbone bonded to three fatty acids. The fatty acids are further distinguished into saturated fatty acids, which have no double bonds in the fatty acid chain, and monounsaturated fatty acids (one double bond), and polyunsaturated fatty acids (multiple).

Soybeans are rich in Linoleic (18:2) and Linolenic (18:3) fatty acids, both of which are essential to the human diet and cannot be synthesized in our bodies.[131,148] These polyunsaturated fatty acids contain double bonds susceptible to oxidation and the generation of off-flavours, but this is not a major problem in natto production since the soybean cells are kept intact, preventing the introduction of oxygen and enzymes into direct contact with the oil bodies.

In addition to triglycerides, soybeans contain a very small amount of free fatty acids.

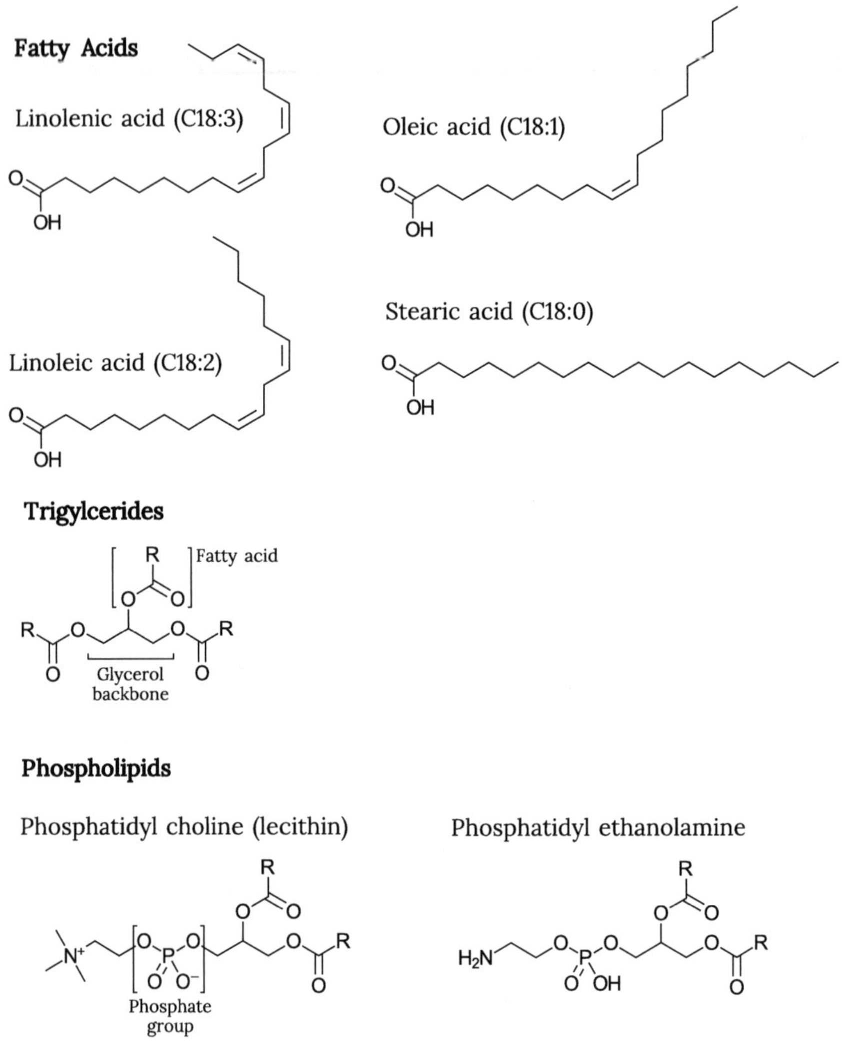

Figure V.17 Selected lipids found in soybeans

Phospholipids

Phospholipids are the next most abundant lipid in soybeans, and comprise 1.5%–2.5% of the total lipids. The basis of all phospholipids is phosphatidic acid, which contains a glycerol backbone, two fatty acids, and a phosphate group. The phosphate group bonds to choline, ethanolamine, and inositol to create phosphatidyl choline (lecithin), phosphatidyl ethanolamine, and phosphatidyl inositol, respectively. Other phospholipids are also present in trace amounts.[131,148]

These phospholipids are polar, are good emulsifying agents, and are found abundantly in the cell membranes.

Soybean Lipids in Natto

The lipids of soybeans remain mostly unchanged during the natto fermentation process. The steaming process has been shown to slightly reduce lipids, mostly as a reduction in sterols and free fatty acids.[149] Although *B. subtilis* is able to produce lipases,[150] it does not appear to do so during natto-making.

2.6 Minor Components

There are a few minor components that affect the overall nutrition of the soybean. Notably, these include phytate, isoflavones, minerals, and vitamins. The amounts of each of these compounds varies by cultivar and growing conditions, and the health effects of some of these components is discussed in Chapter VI.

Phytate

Phytic acid is a molecule consisting of six phosphates bonded to a central molecule of inositol. It acts as a store of phosphorus for the soybean seed. In water, the phosphate groups are negatively charged and bond with metal ions like Calcium, Magnesium, Zinc, and Iron, to form poorly soluble metal phytates. Humans lack the enzyme phytase required to break down phytic acid, so it is regarded as an anti-nutrient, although *in vivo* study results vary and it may confer some health benefits.[131,151]

In natto making, soaking reduces phytate content by between 23–30 percent.[152] The concentration of phytate during the fermentation process is not well-studied. *B. subtilis* can produce phytate,[153] and in other food fermentation processes, the concentration of phytates is reduced.[154]

Flavonoids: Isoflavones and Anthocyanins

Isoflavones are a class of molecules that have a molecular structure consisting of two benzyl rings joined by a three-carbon bridge, which is itself sometimes formed into a pyran ring. These molecules belong to a class of compounds known as flavonoids, which are an incredibly diverse class of polyphenols produced by plants for a variety of physiological functions. Legumes use flavonoids to help nitrogen-fixing *Rhizobia* form root nodules.[131,151]

Soybeans are uniquely rich in isoflavones, which are separated into two groups, aglycones (daidzein, genistein, and glycitein) and glucosides. Glucosides are just aglycones with another sugar group bonded to one of the benzyl rings. Additional functional groups (acetyl–, malonyl–, and succinyl–) on that sugar group form more complex glucosides.[131,151]

During the natto making process, these isoflavones convert between multiple forms. Soaking has a very minor effect on isoflavones, but steaming converts most malonyl glucosides into their base glucosides, removing the malonyl group. Subsequent fermentation converts these glucosides into succinyl glucosides, as the metabolism of *B. subtilis* adds a succinyl group.[155,156] These succinylated glucosides are also biologically active.

A detailed study of the isoflavone content throughout the natto production process has been published, with process streams sampled from a natto factory. The isoflavone contents at each step of the natto process, as well as the solids content of each stream, for both whole grain (*tsubu natto*) and *hikiwari natto* are presented on the following pages.

Isoflavone contents are highest in green soybeans, followed by yellow and black soybeans with a similar amount, and brown soybeans having the lowest. The exact amounts vary with cultivar and growing conditions. Anthocyanins are only present in black soybeans primarily as cyanidin-3-O-glucoside, delphinidine-3-O-glucoside, and petunidin-3-O-glucoside.[157]

Table V.6 Isoflavone content throughout tsubu natto manufacturing process.[156]

(mg/100g dry wt.)	**Aglycones**		**Glucosides**	
Material	Daidzein	Genistein	Daidzin	Genistin
Raw soybeans	0.8	n/d	23.2	12.4
Soaked soybeans	1.8	n/d	29.7	12.7
Soaking water	n/d	n/d	n/d	n/d
Steamed soybeans	2.8	n/d	151.4	114.9
Steaming broth	3.4	n/d	221.2	110.7
Natto	1.3	n/d	111.4	67.8

(mg/100g dry wt.)	**Malonyl glucosides**		**Succinyl glucosides**	
Material	Malonyl daidzin	Malonyl genistin	Succinyl daidzin	Succinyl genistin
Raw soybeans	153.0	153.4	n/d	n/d
Soaked soybeans	142.5	159.5	n/d	n/d
Soaking water	n/d	n/d	n/d	n/d
Steamed soybeans	n/d	10.7	n/d	n/d
Steaming broth	38.1	25.1	n/d	n/d
Natto	10.3	8.3	94.1	71.3

Material	**Original solids (%)**	**Dry matter (%)**	**Total isoflavones (mg/100g TS)**	**Total isoflavones (mg)**
Raw soybeans	100.0	90.2	342.7	342.7
Soaked soybeans	99.7	40.4	346.2	345.2
Soaking water	0.3	0.2	0.0	0.0
Steamed soybeans	99.4	39.4	279.8	278.1
Steaming broth	0.3	15.7	398.5	1.2
Natto	93.1	37.7	364.5	339.3

Note 1: n/d = not detected; <0.5mg isoflavones/100g. TS = total solids

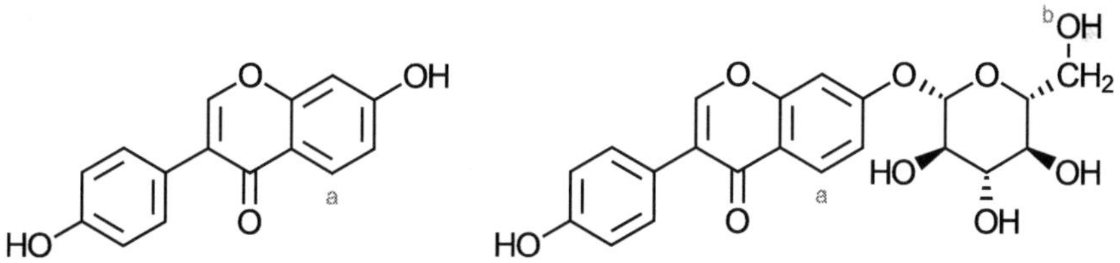

Daidzein　　Daidzin

Figure V.18 Daidzein (left, Aglycone) and Daidzin (right, Glucoside)
Genistein and Genistin contain a hydroxyl (–OH) group at position 'a'
Malonyl and succinyl glucosides contain their respective functional groups at 'b'

Table V.7 Isoflavone content throughout hikiwari natto manufacturing process.[156]

(mg/100g dry wt.)	**Aglycones**		**Glucosides**	
Material	Daidzein	Genistein	Daidzin	Genistin
Raw soybeans	1.4	1.1	25.1	20.6
Cracked soybeans	1.5	1.3	19.2	20.1
Powder (waste)	1.7	1.1	22.6	18.9
Hulls	n/d	n/d	10.7	1.9
Soaked soybeans	7.9	8.3	14.7	11.5
Soaking water	n/d	n/d	n/d	n/d
Steamed soybeans	13.8	5.4	49.0	63.8
Steaming broth	11.9	8.9	78.1	56.0
Natto	8.9	9.5	45.6	52.3
(mg/100g dry wt.)	**Malonyl glucosides**		**Succinyl glucosides**	
Material	Malonyl daidzin	Malonyl genistin	Succinyl daidzin	Succinyl genistin
Raw soybeans	57.7	93.2	n/d	n/d
Cracked soybeans	44.6	91.7	n/d	n/d
Powder (waste)	99.4	79.5	n/d	n/d
Hulls	17.7	9.3	n/d	n/d
Soaked soybeans	45.4	103.5	n/d	n/d
Soaking water	n/d	n/d	n/d	n/d
Steamed soybeans	19.7	40.0	n/d	n/d
Steaming broth	160.6	232.4	n/d	n/d
Natto	16.3	34.4	20.7	25.8
Material	**Original solids (%)**	**Dry matter (%)**	**Total isoflavones (mg/100g TS)**	**Total isoflavones (mg)**
Raw soybeans	100.0	87.1	199.2	199.2
Cracked soybeans	84.8	89.7	178.4	155.8
Powder (waste)	8.4	90.6	223.3	19.6
Hulls	6.3	87.3	39.6	2.5
Soaked soybeans	81.4	39.1	191.3	155.7
Soaking water	3.4	0.21*	n/d	n/d
Steamed soybeans	81.1	41.2	191.7	115.5
Steaming broth	0.3	5.4	547.9	2.0
Natto	78.7	42.0	213.5	168.1

*Number likely an error from original paper; a value of 2.1% is expected here

Minerals and Ash

The soybean's ash content is measured by incinerating it at 550°C, which leaves a flaky grey residue that comprises approximately 5% of the soybean's dry weight. Ash content approximates the soybean's inorganic matter. Physically it contains minerals, and their oxide forms, depending on the mineral. Soybeans are rich in minerals, at its mineral content is detailed in Chapter VI. The mineral content is reduced during soaking and steaming, where some soluble minerals are leached out into the soaking water and steaming broth, respectively.[131,151] Mineral content stays the same during fermentation, during which minerals may convert between different chemical forms but the amount is unchanged.[129]

Vitamins

Soybeans naturally contain a mixture of all essential vitamins except for the fat-soluble vitamin D, and vitamins C and B_{12}. During growth, *B. subtilis* will synthesize vitamins K (in the form of Vitamin K_2 as menaquinone-7), B_2 (riboflavin) and B_3 (niacin), all of which are essential to the basic metabolic processes of *B. subtilis.* As discussed in Chapter IV, Vitamin K_2 is an essential part of the electron transport chain in the respiration of *B. subtilis*, and natto is the richest known food source of this vitamin. Its significance and concentrations will be examined in Chapter VI. On the other hand, raw soybeans contain a small amount of Vitamin A, but this vitamin is not detected after steaming and fermentation.[129]

Long Term Storage of Natto

Natto should be consumed before the expiry date, which is typically 9 days after fermentation or quality evaluation, provided that the natto is refrigerated. However, tests have been conducted on the long term storage of natto under different conditions.

Since natto is not pasteurized, it remains biologically active. During extended storage, bacterial counts generally stay the same, but both ammonia and free amino acids will increase as the enzymes in natto continue to break down proteins. For the same reason, the hardness of natto will decrease, and the colour will darken over time. These effects are accelerated the higher the storage temperature.[158]

Continued storage will liberate more and more amino acids. Eventually, the amino acid tyrosine will precipitate out of the natto, forming small white crystals. Extended storage will also lead to the formation of struvite, which is a mineral of magnesium, ammonia, and phosphate, as the breakdown products of natto precipitate out of the liquid phase. These effects are amplified by the loss of moisture during storage.[159]

2.7 Overall Mass Balance & Waste Streams

Surprisingly little information is available on the overall mass balance of the natto-making process. General rules of thumb are that 1kg of raw soybeans will yield 2kg of natto product, and that 1kg of raw soybeans, when soaked, will yield 2.2kg of soaked beans (at a 2.2 water absorption ratio).

Actual measured and recorded data on the material streams of a natto factory are scarce in the public domain. The most detailed mass balance can be found in the study of isoflavone content (see Section 2.6 above).

In this study, the authors presented the solids content of each material stream as a percentage of the original soybean's solid content, as well as the measured dry matter. These two parameters allow us to calculate the total weight, solids, and water content of material streams for the whole grain (*tsubu natto*) and *hikiwari natto* manufacturing processes.

Table V.8 Mass balance of an example tsubu natto manufacturing process

Material Stream	Original Solids (%)	Total Solids (%)	Total mass (g)	Solids (g)	Water (g)
Raw soybeans	100.0	90.2	100.0	90.2	9.8
Soaked soybeans	99.7	40.4	222.6	89.9	132.7
Soaking water	0.3	0.2	128.9	0.3	128.6
Steamed soybeans	99.4	39.4	227.6	89.7	137.9
Steaming broth	0.3	15.7	1.7	0.3	1.5
Natto	93.1	37.7	222.7	84.0	138.8

Table V.9 Mass balance of an example hikiwari natto manufacturing process

Material Stream	Original Solids (%)	Total Solids (%)	Total mass (g)	Solids (g)	Water (g)
Raw soybeans	100.0	87.1	100.0	87.1	12.9
Cracked soybeans	84.8	89.7	82.3	73.9	8.5
Powder (waste)	8.4	90.6	8.1	7.3	0.8
Hulls	6.3	87.3	6.3	5.5	0.8
Soaked soybeans	81.4	39.1	181.3	70.9	110.4
Soaking liquid	3.4	2.1*	141.0	3.0	138.1
Steamed soybeans	81.1	41.2	171.5	70.6	100.8
Steaming broth	0.3	5.4	4.8	0.3	4.6
Natto	78.7	42.0	163.2	68.5	94.7

*Original paper states a value of 0.21%; however, a value of 2.1% is expected here

This data is taken from a single factory and should not be regarded as 'standard'. Different results would be expected with different soybeans and starting compositions. According to the mass balance of this factory, 100g of small grain soybeans would yield 222.7g of natto, while 100g of large grain soybeans would only yield 163.2g of *hikiwari natto*, since some mass is lost as powder and hulls during the cracking process.

During fermentation, whole grain natto loses about 6% of its solids content as *Bacillus subtilis* consumes portions of the soybean, while this value is only around 3% for *hikiwari natto*. The metabolism of sugars releases some water, while some water is lost due to evaporation. Overall, water content stays mostly the same for both types of natto. In this study, natto was sampled 2 days after end of fermentation for whole grain natto, and 3 days after end of fermentation for *hikiwari natto*, so any moisture lost during refrigeration is also accounted for.

In a different paper, 1000g of steamed soybeans with a total solids of 39.1% yielded 890g of natto with a total solids content of 39.2%, totalling about 10.7% loss of solids during the natto-making process.[144] These numbers also indicate that moisture loss and solids loss occur at approximately the same rate, since percentage total solids content is unchanged.

The steaming broth for whole grain natto is surprisingly concentrated at 15.7% solids. In another paper examining the feasibility of recovering the steaming liquid, the nascent liquid was measured at 12.5% solids (as well as a biological oxygen demand (BOD) of 93000ppm, chemical oxygen demand (COD) of 74000ppm, 5.9% sugars and 1.03% protein), so this value appears to be accurate.[160] Nevertheless, steaming only results in solids losses totalling 0.3% of the original soybean's solids.

In the *hikiwari natto* process, significantly more solids are leached out in the soaking process, which is expected since the beans are cracked open. However, the factory from which this data is collected doesn't appear to employ the 'limited soaking' method for *hikiwari natto* described in Chapter II.1.6.

3. Aroma of Natto

The aroma of natto is hard to describe. In the book 'Nose Dive' by Harold McGee, the author describes natto as "an aroma that combines cocoa and coffee qualities with sweaty and sulfidic notes and sometimes ammonia, along with toasty pyrazines".[161] This matches most consumers' tasting experience, with common tasting notes including butter, toasted notes, coffee, mushroom, fruit, and cheese. Condiments such as mustard and *tare* affect the aroma, and anecdotally, the taste of the condiments can be more pronounced than the natto itself.

Aroma compounds can come from the soybean, a metabolite of the natto-making process, or both. Some components of the steamed soybean's aroma are also reduced or lost during fermentation.[147]

Literature studying the aromas of natto rely heavily on gas chromatography (GC), often combined with other analytical techniques such as in gas chromatography mass spectrometry (GC-MS), or by simply routing the output of a GC to someone's nose (GC-sniffing, or GC-olfactometry). In GC, volatile molecules are separated using a carrier gas that moves the sample through a column. Depending on the type of column, individual compounds are separated based on their physical and chemical characteristics, with lighter, more volatile components being eluted through the column first. The output of the column can then be further analyzed with one of the methods above.

Over 100 components have been identified with this method,[162] and collectively they form the complex aroma of natto.

Table V.10 Selected odour compounds changed or formed during natto fermentation.[147]

Compound	Selected Odour Notes	Change during fermentation
Maltol	Caramel, butterscotch	Decreases
1-octen-3-ol	Mushroom, earthy	Minor change
Hexanal	Woody, vegetable	Decreases
β-damascenone	Woody, floral	Minor change
Phenylacetaldehyde	Green, floral	Decreases
2-methylbutyl acetate	Fruity, banana	Decreases
Hexyl acetate	Fruity, apple	Decreases
3-(methylthio)-propanal	Musty tomato, potato	Decreases
(-)-limonene	Citrus, pine	Increases
1-hexanol	Herbal, pungent	Increases
γ-terpinene	Citrus, lime	Increases
α-terpinene	Citrus, woody	Increases
Methylpyrazine	Nutty, cocoa, roasted	Minor change
2,5-dimethylpyrazine	Musty, potato, cocoa	Created
Trimethylpyrazine	Nutty, chocolate, coffee	Created
Isovaleric acid	Cheesy, acidic, pungent	Created
Isobutyric acid	Cheesy, acidic, rancid	Created
Guaiacol	Smoky, woody	Created

Some of the broader trends are as follows.

1. **Terpenes**, including limonene, γ-terpene, and α-terpene, are present in the cooked soybeans and increase during fermentation. They impart piney and flowery notes to the natto.
2. **Pyrazines**, including 2,5-dimetyhlpyrazine, trimethylpyrazine, and methylpyrazine, increase throughout the process of fermentation, with the latter being also present in cooked soybeans. These compounds impart an earthy, nutty, roasted aroma.
3. **Branched chain fatty acids** (BCFAs) including isovaleric and isobutyric acid, increase through fermentation. These compounds

impart a cheesy, sweaty aroma. BCFAs are reduced when using low-odour strains of *B. subtilis* (see Chapter IV.7.1).

4. The compound **guaiacol** increases during fermentation. This compound imparts a 'smoky' aroma and is associated with roasted coffee.

All natto textbooks highlight the importance of refrigerated aging after fermentation, which is known to allow 'off-odours' and ammonia to volatilize.

Note that the results from the table above are from a single Japanese study. Other studies have identified different compounds at different concentrations.

Odour notes are adapted from an interesting web resource called 'The Good Scents Company', which compiles perceived odour notes from a variety of chemical suppliers and resources.[163]

Overall, the aroma profile of steamed soybeans drastically changes during the fermentation process, and the wide range of compounds associated with natto means that it can be difficult to describe the aroma itself, other than being 'natto-like'. The science of smell and odour is inherently subjective and difficult to quantify.

Summary

The soybean, *Glycine max*, is an incredibly important agricultural commodity due to its high protein and fat content, and high agricultural yield. During natto fermentation, the main transformations that happen to the soybean are the digestion of soluble sugars by *B. subtilis*, and the partial digestion of proteins into its constituent amino acids and ammonia. Both processes yield new metabolites, including vitamins and biofilm components, and heat. Most of the soybean's other components, like insoluble fibres and lipids, are largely unchanged. These conversions affect natto's nutritional and physiological properties when compared to raw soybeans, which will be discussed in the next chapter.

Further Reading

Whole textbooks have been dedicated to soybeans due to their agricultural significance. Any of the following textbooks are valuable resources that detail everything about soybeans, including its detailed biology, and strain breeding, which were excluded from this chapter:

- **Practical handbook of soybean processing and utilization** by David R. Erickson.[164]
- **Soybeans: chemistry, technology and utilization** by Keshun Liu.[165]
- **Soybeans: chemistry, production, processing, and utilization** by Lawrence Alan Johnson, Pamela J. White, Richard Galloway.[166]

For good references on food chemistry, refer to any of the following:

- **Food chemistry** by H.-D. Belitz, W. Grosch, Peter Schieberle.[167]
- **Fennema's food chemistry, 5e** by Srinivasan Damodaran, Kirk L. Parkin.[168]

For an in-depth, but approachable book on the science of scent, I recommend the following book by a legendary food author:

- **Nose Dive: a field guide to the world's smells** by Harold McGee.[161]

Lastly, the Good Scents Company (tgsc) has a huge online database of scent compounds.

- **www.thegoodscentscompany.com**

VI. Nutrition and Physiological Effects of Natto

Natto producers and avid consumers alike are quick to regurgitate a long list of health benefits from regularly consuming this package of sticky beans and bacteria. They'll claim that it's nutritious, lowers blood pressure, lowers cholesterol, prevents osteoporosis, improves bone health, is good for menopausal women, is good for everyone, is probiotic, is prebiotic, improves digestibility, is anti-aging, anti-cancer, anti-allergy, stimulates the immune system, improves skin health, and a whole host of others.

The reality is that most of these benefits still need to be studied further before claims can be made with a high degree of confidence. The science of nutrition and physiology, arguably more so than any other field, is prone to bias, statistical error, and the phrase "correlation does not imply causation". These issues are occasionally amplified by low quality studies and low quality journalism, headlining with the phrase: "studies have shown…".

It can also be easy to jump to conclusions based on an overly simplified understanding of science. Consider the claim: "natto is full of bacteria, so it is an excellent probiotic". While it's true that natto contains both live bacteria and endospores, as of writing this chapter, we don't fully understand if and how the bacterium confers benefits to gut health, or if certain strains of *B. subtilis* are better at doing so than others.

The gold standard for the study of human nutrition and physiology is the randomized controlled trial (RCT).[169] When a health claim is made about a particular food or substance, the ideal test scenario is through a random group of subjects with sufficient size, and a properly designed protocol, eliminating as many convoluting variables as possible. Furthermore, the mechanism behind the proposed health benefit should be theoretically valid or elucidated. Anything short of this standard must be taken with a grain of salt. Unfortunately, these trials can be prohibitively expensive, so often times we're left with cheaper animal trials, epidemiological studies, diet history questionnaires (DHQ), food frequency questionnaires (FFQ), other surveys, and *in vitro* studies. Even when they are completed, RCTs are often hampered by short time periods, poorly designed experiments, or marred with conflicts of interest.

Most of the purported health benefits of natto are related to cardiovascular diseases (CVDs), which according to the World Health Organization account for the top three leading causes of death globally. Other possible health benefits relate to calcium uptake and transport, cancer prevention, the prevention of osteoporosis, and the improvement of gut microbiota.

Overall, natto is generally regarded as safe and nutritious, even when ignoring all these 'bonus' health benefits. The fact remains that you are eating a whole grain, plant-based source of protein.

Natto is also unique when compared to other soy-based foods. For example, unlike tofu and tofu products (which have all its fibre removed), it is still a whole grain soy food. Unlike *shoyu*, *miso*, and *hama natto*, it contains no added sodium, so it can be consumed as a staple rather than a seasoning. Unlike *tempeh*, it is typically consumed raw, making it more convenient to eat, and allowing it to still confer health benefits from heat-sensitive biological compounds.

In this chapter we will examine all the nutritional factors of natto, as well as many of the claimed benefits, while considering some of the experiments and studies that were conducted in each case, where applicable.

Disclaimer

Always consult your health care provider before making any health care decisions or for guidance about a specific medical condition.

1. Nutritional Value of Natto

1.1 Basic Macronutrient Data

Natto inherits almost all the nutritional benefits of eating whole grain soybeans, including its high protein, fat (especially unsaturated and polyunsaturated fatty acids), and fibre content. The macronutrient data of *itohiki natto* and *hikiwari natto*, as recorded by the Standard Tables of Food Composition in Japan 2020 (8th revision), are provided below.[129]

Table VI.1 Macronutrient data of itohiki natto and hikiwari natto, per 100g wet.

Parameter	Units	Itohiki natto	Hikiwari natto
Energy	kcal	190	185
Water	g	59.5	60.9
Protein (by amino acid analysis)	g	14.5	15.1
Protein (by N content)	g	16.5	16.6
Triglycerides	g	9.7	9.7
Cholesterol	mg	trace	0
Lipids	g	10	10
Available carbohydrates (monosaccharide equivalents)	g	0.3	0.2
Available carbohydrates (mass meter)	g	0.3	0.2
Available carbohydrates (subtraction method)	g	7.7	6.4
Dietary fibre	g	6.7	5.9
Sugar Alcohol	g	-	-
Total carbohydrates	g	12.1	10.5
Organic Acids	g	-	-
Ash	g	1.9	2.0

Proteins

The amino acid profile of soybeans its mostly unchanged when converted into natto, and soybeans remain one of the few legumes containing a significant amount of methionine, which is usually a 'limiting' amino acid in other legumes.[170] However, in a well-balanced diet containing a variety of plant protein sources, the concept of limiting amino acids is not important.[171] Natto contains 41% protein on a dry weight basis, and a single 50g package of natto contains 8.2g of protein. Even accounting for its slightly inferior protein digestibility, this is more than a 50g hard-boiled egg which contains 6.3g of protein (a standard large, hard-boiled egg is also 50g wet).

Studies on the digestibility of natto protein have mixed results. Many researchers hypothesize that natto should be more digestible than normal soy protein (such as from tofu, boiled soybeans, etc.), since the proteins have been partially digested by the feeding proteases of *B. subtilis*.[172] Results are mixed: different studies have shown that natto possesses roughly the same protein digestibility as boiled soybeans, but less than processed soy products such as tofu. The conclusion of the Science and Technology Agency of Japan is that the digestion-absorption rate of natto proteins is around 90%.[173]

Fats and Lipids

Natto contains a nearly identical lipid profile to boiled soybeans, since lipids are largely unchanged during the fermentation process.[149] The lipid content of natto is 24.7% on a dry weight basis. Of the triglycerides, the most abundant fats are polyunsaturated fatty acids, followed by monounsaturated fatty acids and saturated fatty acids. Natto contains the essential fatty acids linoleic acid and linolenic acid. As a plant and bacterium-based food, natto contains no cholesterol.

Carbohydrates

Natto contains less free sugar than boiled soybeans, and a low amount of raffinose family oligosaccharides (RFOs) since these have been digested or converted by *B. subtilis* into other substances (see Chapter V). This reduces flatulence, since these RFOs would otherwise be left intact for bacteria in the large intestine, causing the release of gases.[138] However, these RFOs are considered good for gut bacteria, so the loss of RFOs may be considered a disadvantage of natto compared to boiled soybeans. Nevertheless, natto still confers other potential benefits to the gut microbiome through levan and other substances. Furthermore, natto retains all of its insoluble fibre from boiled soybeans, which are also beneficial for general health and the gut microbiome.[174]

Overall, natto is a food product with a desirable nutritional profile, and a significant amount of protein and unsaturated fatty acids, while still containing high amounts of fibre and low amounts of free sugars.

1.2 Vitamins and Minerals

Natto retains most of the minerals present in soybeans, and alters its vitamin profile (see Chapter V), adding a large quantity of Vitamin K_2. This vitamin is synthesized by *B. subtilis* and implicates natto in all studies of the physiological benefits of this vitamin.

Table VI.2 Mineral content of itohiki natto and hikiwari natto, per 100g wet

Parameter	Units	Itohiki natto	Hikiwari natto
Sodium	mg	2	2
Potassium	mg	660	700
Calcium	mg	90	59
Magnesium	mg	100	88
Phosphorus	mg	190	250
Iron	mg	3.3	2.6
Zinc	mg	1.9	1.3
Copper	mg	0.61	0.43
Manganese	mg	-	1
Iodine	µg	trace	-
Selenium	µg	16	-
Chromium	µg	1	-
Molybdenum	µg	290	-

Table VI.3 Vitamin content of itohiki natto and hikiwari natto, per 100g wet

Parameter	Units	Itohiki natto	Hikiwari natto
Vitamin A Compounds			
Retinol	µg	0	0
alpha-carotene	µg	-	0
beta-carotene	µg	-	0
beta-cryptoxanthin	µg	-	0
beta-carotene equivalent	µg	0	0
retinol activity equivalent	µg	0	0

Parameter	Units	Itohiki natto	Hikiwari natto
Vitamin D	µg	0	0
Vitamin E Compounds			
alpha-tocoperol	mg	0.5	0.8
beta-tocoperol	mg	0.2	0.3
gamma-tocoperol	mg	5.9	9
delta-tocoperol	mg	3.3	5.4
Vitamin K	µg	600	930
Thiamin (B1)	mg	0.07	0.14
Riboflavin (B2)	mg	0.56	0.36
Niacin	mg	1.1	0.9
Niacin Equivalent	mg	5.2	5
Vitamin B6	mg	0.24	0.29
Vitamin B12	µg	trace	trace
Folic acid	µg	120	110
Pantothenic acid	mg	3.6	4.28
Biotin	µg	18	-
Vitamin C	mg	trace	trace

2. Physiological Effects Natto

Japanese natto producers will make many claims about the health benefits of natto as part of their marketing. Of course, I do not think that the natto industry is malignant, but there may be an inherent bias and potential conflict of interest when health claims are made by the same group producing and selling the food in question.

Many of the mechanisms behind the physiological effects of natto are yet to be fully elucidated. For some of these benefits, only a small number of studies have been completed.

Furthermore, nutritional science is constantly evolving. The contents of this chapter were written in December of 2022, and as of writing, papers as recent as 2021–2022 still conflict with the findings of prior research. Thus, each health claim described in this section may change as more studies are completed.

2.1 Vitamin K2 (Menaquinone)

Vitamin K was first discovered by Henrik Dam in 1929, when he found that feeding chicks with food that had been extracted with nonpolar solvents caused their blood to clot slowly. He concluded that there must be a fat-soluble substance involved in blood clotting and named it Vitamin K after the Danish word 'Koagulans'.[175] We now know that Vitamin K comes in three major forms (vitamers), and has several physiological functions beyond blood clotting.

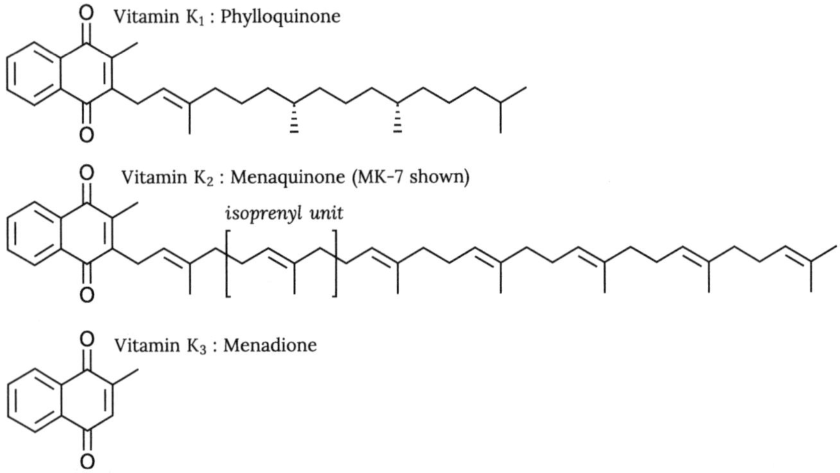

Figure VI.1 Forms of Vitamin K

Vitamin K_1 (Phylloquinone) is found in leafy green vegetables, since phylloquinone is involved in the electron transport chain in chloroplasts.

Vitamin K_2 (Menaquinone) is synthesized by many bacteria for use in their respiratory electron transport chains, including *Bacillus* (see Chapter IV.5.2). Menaquinones are further denoted as MK–n, where 'n' is the number of isoprenoid units on its side chain. The form created by *Bacillus subtilis* is MK–7.

Vitamin K_3 (Menadione) is synthetic and no longer used due to side effects. Chemically, it's the functional group of Vitamins K_1 and K_2, without the hydrophobic side chain.

The two natural forms of Vitamin K are physiologically different. Vitamin K_1 is comparatively difficult to absorb, and sources from leafy green vegetables are considered to have less than 15–20% bioavailability. By comparison, Vitamin K_2 is much more readily absorbed and stored abundantly in the liver. The body's pool of K_1 is also much lower than K_2, and the turnover of K_1 is much quicker than K_2.[175]

Natto is implicated in studies of Vitamin K_2 because it is the single largest food source of the vitamin by an order of magnitude, even when comparing between single servings of different sizes, or side-by-side comparisons of the same wet weight. 100g of natto contains between 600µg–930µg of Vitamin K_2 as MK-7,[129] but this estimate is conservative and varies based on fermentation conditions. Concentrations as high as 1750µg/100g have been reported.[176]

Carboxyglutamic Acid (Gla)

Vitamin K is involved in a class of post-translationally modified proteins, called Vitamin K Dependent Proteins (VKDPs). All these proteins contain a 'Gla domain' with several amino acids called carboxyglutamic acid. This amino acid is not one of the 20 amino acids used in the process of protein synthesis by translation in the human body. Instead, it is formed by the post-translational modification of glutamic acid with the enzyme γ-carboxylase.

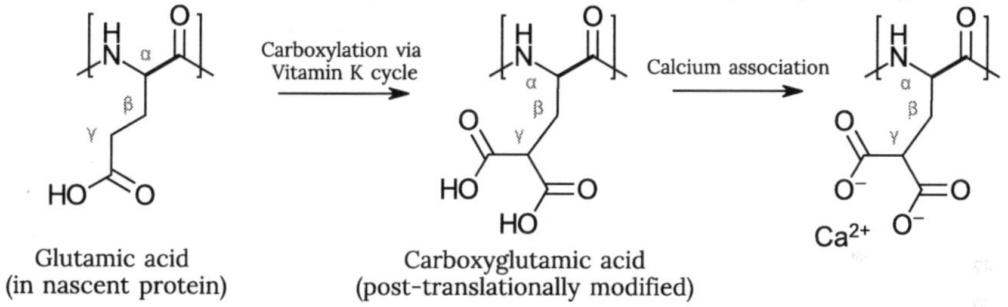

Figure VI.2 Formation of carboxyglutamic acid

Through a cyclic pathway involving Vitamin K, this carboxylation reaction adds an additional carboxyl group to glutamic acid (Glu), and the resulting amino acid (Gla) contains two carboxyl groups and gives the protein a strong affinity for calcium ions (Ca^{2+}).[175]

These calcium-binding VKDPs are found in many proteins where calcium ions are involved, including hemostasis-related plasma proteins, bone, and tissue calcification.[175] Some of the well-studied VKDPs will be presented in this section, but there are other VKDPs and their functions are still being actively researched.

Hemostasis-Related Plasma Proteins

VKDPs are involved in the process of hemostasis, which is your body's solution to prevent bleeding out and repairing wounds in the event of an injury.

Proteins involved in the coagulation process, including factors VII, IX, X, and prothrombin (factor II), are all VKDPs. These proteins form the coagulation cascade, which is a series of reactions that form a cross-linked fibrin mesh during secondary hemostasis, trapping red blood cells and creating a plug known as a blood clot (primary hemostasis is a temporary plug using your platelets).

On the other hand, a separate system of anticoagulants clears the blood clot after the wound has healed. These include proteins C, S and Z, which are also VKDPs.[175]

Bone Health and Osteocalcin

VKDPs are also involved in the formation and regulation of calcium and calcified tissue. The protein osteocalcin (OC) has been shown to be involved in the alignment of biological apatite in bone (its primary mineral component), which increases its strength.[177] OC exists in two forms, its uncarboxylated (ucOC) and carboxylated (cOC) forms, the latter of which is a VKDP. ucOC may be implicated as a hormone, but currently, its exact function and mechanisms are still being investigated.

Blood Clots and Cardiovascular Disease

Another protein, the matrix Gla protein (MGP), is synthesized in cartilage and other soft tissues, and is essential to prevent vascular calcification.[175] Without it, calcium builds up in extracellular matrix (ECM) in blood vessels, causing them to harden.[178] In severe atherosclerosis, the build up of plaque and calcium deposits along the walls of the blood vessels, called an atheroma, can eventually rupture and cause create a blood clot within the blood vessel, leading to a condition known as ischemia. Thus, MGP is considered essential for vascular health.

Other Vitamin K Dependent Proteins

There is a long list of other potentially important Vitamin K dependent proteins, but their function has not been fully elucidated.

Vitamin K Dosage

Despite the importance of Vitamin K in the body, and natto's exceptionally high concentration of the vitamin, there's no conclusive evidence that high dosages of Vitamin K are beneficial. Clinical Vitamin K deficiency is exceedingly rare, and some amount of Vitamin K is also produced by your gut bacteria and absorbed.[179] Vitamin K is often given to neonates (newborns) who are otherwise susceptible to Vitamin K deficiency bleeding (VKDB) since their blood does not have enough Vitamin K to facilitate clotting, and they have a relatively sterile gut.

The Daily Recommended Intake (DRI) of a vitamin is typically calculated from the Estimated Average Requirement (EAR), the median daily amount that meets the requirements of 50% of healthy individuals. However, for Vitamin K, there is insufficient data to support the calculation of the EAR, so instead the Adequate Intake (AI) parameter is used, which is an estimate based on experimental observations.

For Vitamin K, the AI values are 120μg/d for adult males and 90μg/d for adult females. There is no indication that large amounts of Vitamin K are toxic, so the current DRI sets no tolerable upper intake limit for Vitamin K.[175]

A single 50g pack of natto can supply roughly three times the current DRI for Vitamin K, and in its more readily absorbed and longer lasting form Vitamin K_2 (as MK–7). Current research does not provide conclusive evidence that the high doses of Vitamin K_2 found in natto have any additional benefits besides easily meeting the DRI.[175]

2.2 Nattokinase (subtilisin NAT)

Nattokinase was first discovered and described by Hiroyuki Sumi, a Japanese Researcher working at the Chicago Medical School in 1980, who discovered that natto could dissolve fibrin on a petri dish.[90,180] Working with other Japanese researchers, they were able to extract the active substance and demonstrate that 1g of wet natto had about the same fibrinolytic activity as about 40 Casein Units (CU) of plasmin or 1600 International Units (IU) of urokinase, which is significant enough to warrant further investigation into its potential as a therapy for cardiovascular diseases.[10] They named this substance nattokinase, after the words 'natto' and 'urokinase'.

Characterization of Nattokinase

Sumi's research team performed basic characterization of nattokinase, noting that the enzyme is stable with freezing and thawing, but is slowly inactivated above 60°C. It is stable between pH values of 7–12, but labile below a pH of 5.0, although this could be stabilized with some additional substances (including boiled rice extract, boiled meat extract, albumin, and mucin).

Aside from being able to directly digestion fibrin, nattokinase has also been demonstrated to activate tissue plasminogen activator (tPA) and activate pro-urokinase into urokinase, both of which participate in conversion of plasminogen to plasmin, the major human enzyme responsible for breaking down blood clots. Natto may also contain a pro-urokinase activator that is separate from nattokinase. However, these mechanisms are not well understood.[173]

In subsequent studies, it was discovered that the gene encoding for nattokinase was *AprN*, which is *B. subtilis* var *natto*'s equivalent to the common *Bacillus subtilis* protease, subtilisin E, and it was suggested that the more accurate name for this enzyme would be 'subtilisin NAT'. Regardless, the more approachable name of nattokinase has stuck around.[181] Further work describes the enzyme as a protease consisting of 275 amino acids, and a molecular weight of 27.7kDa.[182]

In vivo (animal) studies

In vivo animal studies for nattokinase started in 1995, when Japanese researchers started testing the efficacy of nattokinase in rats. They found that when injected into the duodenum (the first part of the small intestine), nattokinase could be absorbed into the bloodstream.[183] Further experiments demonstrated that nattokinase had a positive effect on clearing blood clots in rats, for both directly injected and orally administered nattokinase.[184,185]

Human Trials

By now, nattokinase was showing enough promise for trials in human subjects. *In vitro* work demonstrated that on a lab bench, nattokinase could reduce the viscosity and red blood cell aggregation in human blood,[186] but it wasn't clear how nattokinase was transported through oral administration to absorption in the bloodstream, considering that the acidic stomach would easily denature the enzyme.

Only two pharmacokinetic studies of nattokinase in humans have been conducted to date, in 2013 and 2015. In the first, it was shown that orally administered nattokinase peaked in blood around 13.3h (±2.5h) after ingestion,[187] while the second study showed indirectly, through fibrin degradation products, that the activity peaked around 2–8h after oral administration.[188]

The first randomized controlled trial (RCT) was reported in Korea in 2008, where patients with pre-hypertension and stage 1 hypertension (elevated blood pressure) were administered nattokinase capsules (2000 fibrinolytic units (FU), per capsule, around the equivalent of one pack of natto) or a placebo. Over the course of 8 weeks, the subjects given nattokinase showed a mild but statistically significant drop in systolic and diastolic blood pressure.[189]

A second RCT was reported in 2016, this time in North America, and mirrored the first experiment with an 8–week study on subjects with elevated blood pressure. Each subject took one 2000FU capsule every morning for 8 weeks. This experiment also reported a drop in blood pressure, but it was only statistically significant in males.[190]

Researchers noted that both RCTs were relatively short in duration, with small sample sizes. A much larger trial was put together, named the Nattokinase Atherothrombotic Prevention Study (NAPS), this time spanning three years, and studying 265 randomized healthy adults (with normal blood pressure) each administered the standard 2000FU capsule daily to see if it could significantly prevent atherosclerosis. The study showed an insignificant effect on the subjects' blood pressure, suggesting that nattokinase has a null effect on already healthy adults.[191]

All these studies, and others, demonstrate that nattokinase is safe to consume.

Experimental Considerations

The human trials of nattokinase seem to be limited by the design of their experiments. They were primarily designed to study the oral administration of nattokinase extracts as a therapy, rather than unpurified nattokinase within natto itself. Of course, this is understandable since you want to remove as many confounding variables as possible, and whole packages of natto contain much more than just nattokinase. Natto would also be extremely difficult to administer as both a therapy and a placebo: you would have to formulate a natto product that is identical to normal natto, without nattokinase, and then have the control participants consume it daily for the allotted time.

Theoretically, this could be done, but with a lot of difficulty. A 'nattokinase-free' placebo could be developed by fermenting natto with a strain of *B. subtilis* that produces subtilisin E rather than subtilisin NAT, and is otherwise identical. However, administering 50g packages of natto daily would be extremely difficult compared to 100mg pills of nattokinase extract.

In all cases, the nattokinase was provided by Japan Bio Science Laboratory Co., Ltd., a pharmaceutical company that produces and markets the product as NSK–SD, which can then be repackaged for retail by third-party companies. It isn't clear whether the administered nattokinase capsules were designed for release in the duodenum or had any design factors for stomach acid resistance. It's expected that these nattokinase supplements were administered with enteric coatings, which are designed to resist stomach acid and fully dissolve in the more alkaline small intestine.[192]

It also isn't clear how the pharmacokinetics of nattokinase would behave in its unrefined natto form, either standalone or as part of a full meal consisting of rice, soup, and other dishes. We know that the stomach has a fasting pH of 1.0–1.7,[193] which would quickly denature pure nattokinase. However, we also know that during a meal, the stomach pH is quickly brought up to pH 4.5 before being gradually lowered to fasting pH, reaching pH 3.0–3.5 after one hour and pH 1.5–2.0 after two hours.[194] During this time, the stomach continually empties its contents as chyme into the small intestine, with the time needed to empty half the stomach occurring just over one hour after ingestion.[195] Hiroyuki Sumi demonstrated that nattokinase is stabilized by several substances and proteins in food,[90] and it would be safe to assume that nattokinase, when taken as part of a meal, retains some of its efficacy when it reaches the small intestine.

If nattokinase makes it to the small intestine, it can cross into the bloodstream. Although it was previously thought that the small intestine could only absorb smaller amino acids and peptides, we now know that intact proteins, including serine proteases like nattokinase, can be directly absorbed into the bloodstream through several mechanisms.[196]

Only one small study (identified earlier) attempted to quantify nattokinase absorption in the bloodstream of humans, and no complete studies have been done to quantify the amount of intact nattokinase that makes it through various parts of the GI tract and into the bloodstream. Understandably, these parameters are hard to measure and would involve intrusive procedures, but the fact remains that the pharmacokinetics of nattokinase in whole natto are not well-studied. Further work is needed, and for now nattokinase can be considered promising at best.

Other Fibrinolytic Enzymes

Fibrinolytic enzymes found in food are not limited to *Bacillus subtilis* var *natto.* Similar enzymes, including other serine proteases, have been discovered in many other microorganisms, including fungi, bacteria, fermented foodstuffs, and plants. If they prove to be therapeutically useful, and there is a way to administer them safely, any one of them can be developed into a therapy for CVDs.

2.3 Other Potential Physiological Effects

Several other physiological benefits of natto are commonly stated in literature and Japanese natto textbooks. As of writing, these remaining benefits don't have sufficient research or evidence to stake any significant claims. Unlike high-dosage Vitamin K2 and Nattokinase, some of these physiological effects are not unique to natto.

Poly-γ-glutamic acid and Calcium absorption

Poly-γ-glutamic acid (γ-PGA), which is a negatively charged polymer, may improve the solubility and absorption of calcium ions.[197] A 2006 study on rats demonstrated that levels of calcium absorption were significantly increased when diets were supplemented with γ-PGA, and resulted in lower excreted calcium as well as higher bone strength.[198] However, studies on this subject matter are limited, and have only been conducted *in vitro* and *in vivo* in animal studies.

Gut Microflora

Natto is often being cited as a traditional treatment of gastrointestinal upsets through probiotic effects.[173] If natto produces probiotic effects, these effects are not properly understood. *B. subtilis* is a facultative aerobe native to soil environments and would find it difficult to compete in the oxygen-free environment of the large intestine, even if it's capable of anaerobiosis. It's possible that spores germinate, then immediately lyse, and this has a favourable effect on gut bacteria.

Natto may have several prebiotic effects. Limited studies have shown that ingestion of one pack of natto per day (50g per day) for two weeks significantly improved gut *Bifidobacterium*, which is a ubiquitous and beneficial member of the human gut microbiome. This may be a result of consistently consuming a high fibre soy food, rather than from *B. subtilis* itself.[199] Levan, one of the components of natto's mucilage, also supports the growth of gut bacteria.[200] Gut health and the microbiome have gained a lot of attention in recent years, and research is ongoing.

Isoflavones

Soybeans are major sources of isoflavones, which were described in Chapter V.2.6. Natto contains a total of 364.5mg/100g dry weight of isoflavones, or around 73.8mg per 50g pack of natto.[156]

Isoflavones have been implicated in the regulation of estrogen, cancer prevention, use as an antioxidant, lowering of cholesterol, improving calcium absorption, and reducing blood pressure.[201] However, the studies claiming these health benefits have a large variability in dosage levels and experimental design.

There is some concern, especially in the western world, that soy consumption affects male reproductive hormones through isoflavones, but large scale meta-analysis indicate no correlation.[202] Geographically, per capita isoflavone intake in Japan ranges from 26–54 mg/d, whereas in the United States and Europe, it is estimated at 0.5–3mg/d.[203] The difference is due to the greater consumption of soy foods in East Asia.

Saponins

Soybeans are particularly rich in saponins, which are glycosides of a ringed molecule, usually a steroid or triterpene, and named after their foamy appearance when agitated with water. Saponins are antioxidants and may also be involved with anti-cancer effects. Due to their structural similarity to cholesterol, they may lower cholesterol uptake by occupying bile salts.[174]

Biogenic Amines and Polyamines

Natto contains biogenic amines, including tryptamine, β-phenylethylamine, putrescine, cadaverine, histamine, tyramine, spermidine, and spermine.[204]

Biogenic amines are mostly products of amino acids that have undergone enzymatic decarboxylation and other reactions. For example, histamine is formed from the decarboxylation of histidine, and tyramine is formed from the decarboxylation of tyrosine. These compounds are found in significant concentrations in fermented vegetable and dairy products and preserved fatty fishes.[205] Soybeans also contain polyamines that are partially degraded by *B. subtilis* in natto.[173]

Certain biogenic amines are toxic at high concentrations, although exact numbers have not been determined. Maximum recommended amounts of histamine and tyramine must consider the typical serving size of the food containing these biogenic amines. Histamine and tyramine limits in meat and fish have been set to no greater than 100–200mg/kg, although servings of meat and fish are typically larger (around 200g) than serving of natto (50g). Generally, Histamine levels above 500mg/kg or tyramine levels above 1000mg/kg are considered toxic.[205]

Very few studies have evaluated the concentration of biogenic amines in natto. A Korean study of 17 Japanese natto products and 4 Korean natto products found a high variability in each type of biogenic amine, but not enough to cause toxicity at typical serving sizes.[204] However, the presence of biogenic amines may put a limit on the amount of natto that can be consumed in one sitting, although this quantity is not defined and there have been no reported cases of acute toxicity from biogenic amines due to natto consumption.

Natto and Warfarin

Natto, which contains high levels of Vitamin K_2, is antagonistic with the blood-thinning drug warfarin, which interrupts the Vitamin K_2 Gla protein cycle, thus disrupting components of the coagulation cascade. Thus, natto should not be taken with warfarin. Always consult a professional on food-drug interactions when given a prescription.

Natto-induced Late-onset Anaphylaxis

There are recorded cases of natto-induced late-onset anaphylaxis, which is a delayed allergic reaction to the consumption of natto. Symptom onset occurs 5–14 hours after ingestion, and includes rashes, shortness of breath, loss of consciousness, collapse, vomiting, and diarrhea.[206] The exact cause is unknown, but the reaction is IgE-mediated, meaning that the IgE allergy antibodies cause the reaction (these are the same allergic reactions associated with many foods such as milk, eggs, wheat, nuts, etc.) It is noted that natto-induced late-onset anaphylaxis is exceeding rare.

Epidemiological Studies

In one of the largest Japanese epidemiological studies, the Takayama Study,[207] which includes 29,079 subjects spanning 16 years, natto intake was significantly correlated with a reduced risk from cardiovascular diseases. However, large-scale epidemiological studies are prone to confounding variables and these two factors could be a correlation, rather than a causation. These effects may also be a combination of many effects, including nattokinase, Vitamin K2, the act of eating a whole grain soy food, or the displacement of animal proteins with soy proteins in the diet.

Summary

Based on its macronutrient and micronutrient profiles, natto is considered a healthy, whole grain food that happens to be an excellent source of Vitamin K2. Other components of natto may confer additional health or physiological benefits, but all these factors require further study. This includes nattokinase, which is heavily marketed as a health benefit of natto, but only has a handful of limited randomized controlled trials studying its efficacy and pharmacokinetics.

Nevertheless, the future study of natto's physiological effects hold a lot of promise, but natto producers and consumers should exercise some caution when making any health benefit claims. Moreover, the many potential physiological benefits of natto may be synergetic, and should be studied as a whole in addition to studying each substance in isolation.

Further Reading

The science of health and nutrition is one of the most complicated subjects studied in this book, and far beyond my comfort zone or area of expertise. One of the most helpful resources I found on nutrition was:

- **Modern Nutrition in Health and Disease** edited by A. Catharine Ross et al.[208]

The health benefits of natto, including studies of each purported benefit, were covered in the following books in detail.

- **Handbook of Fermented Functional Foods** edited by Edward R. Farnworth. Natto Chapter by Tomohiro Hosoi and Kan Kiuchi.[209]
- **Health Benefits of Fermented Foods and Beverages** edited by Jyoti Prakash Tamang. Natto Chapter by Toshirou Nagai.[173]

VII. Foods Similar to Natto

Natto is not the only food to apply the concept of 'bacteria grown on soybeans'. The soybean's contrasting properties–it's high agronomical efficiency and unappealing taste–means that soybean growing regions of the world have historically adapted methods to transform them into something more delicious. These include processed soybean products like *kinako*, *yuba*, tofu and soybean sprouts, and a large family of fermented soybean products that are often used as flavourings and seasonings.

Depending on the moisture of the soybeans and fermentation conditions, fermentation is usually facilitated by bacteria or fungi. Generally, high moisture favours the growth of bacteria. We can define 'natto-like foods' as non-saline ferments of soybeans *(Glycine max)* using bacteria of the genus *Bacillus* as the primary fermenting microorganism. These bacterial ferments span across the Himalayas, Southeast Asia, and East Asia, and can be eaten directly, mixed with other ingredients, dried, or stewed in curries and soups.

Noteworthy mentions not included in this definition are West Africa's *dawadawa*, which is made from African Locust beans *(Parkia biglobosa)*, and Nigerian *ugba*, made from African Oil beans (*Pentaclethra macrophylla*). While not covered in this chapter, these two foods are examples of fermentation processes adapted for beans that are difficult to consume without fermentation, just like soybeans.

Lower moisture favours fungal ferments, such as Chinese *qu*, Korean *meju*, and Japanese *koji*, which sometimes involve grains such as rice, wheat, and barley. These products are usually mixed with salt after the initial fermentation step, and allowed to mature into seasoning pastes and liquids such as *Jiang*, *Shoyu*, and *Miso*.

Another noteworthy mention is Indonesian *tempeh*, which is a fungal ferment of soybeans with *Rhizopus spp.* used as a primary protein source rather than a seasoning, playing a similar role as natto.

We will also look at foods within Japan that bear the name 'natto', even though they are not 'natto-like foods' by our definition. Learning about these foods gives us more context into the origins of natto.

Figure VII.1 Natto-like foods across Asia

1. Natto-like Foods in East Asia

Korea: Cheonggukjang

Travelling west from Japan, we find Korean *Cheonggukjang* (Korean: 청국장). The traditional process for making *cheonggukjang* is almost identical to natto. Soybeans harvested in the autumn are washed, boiled, cooled, then placed in a warm part of a room: either in straw basket or clay pot covered with straw, then wrapped in a blanket to retain heat.[5] The soybeans are left to ferment for

2–3 days. The bacterium responsible is *Bacillus spp.*, which is found in abundance in dried rice straw.

The fermented soybeans may be mixed with other ingredients such as salt, green onions, garlic, and chili peppers and pressed into cakes to be sold. Its primary use is in a stew (*Cheonggukjang-jjigae*, 청국장찌개), although it can be eaten directly. This mirrors Japanese natto, which is typically eaten directly, and less often used in a soup as *natto-jiru*.

It is possible that Japanese natto originates directly from *cheonggukjang*, and historically, most of Japanese agricultural technology originates from continental Asia, often through modern-day Korea and China.[5] When examining the etymology of *cheonggukjang*, we find the same set of Chinese characters (Korean Hanja, or Japanese Kanji) used interchangeably to describe the family of products that includes Korean *meju*, *jang*, and *kuk*; Japanese *koji*, *miso*, *shoyu*; and Chinese *qū*, *jiang* and *douchi*. Interestingly, the origins and etymology of *cheonggukjang* also include legends about quick-fermented soybeans prepared during war, just like the legend of Minamoto no Yoshiie.[5]

China: *Douchi*

Both *cheonggukjang* and natto can trace their origins to the Chinese food douchi (Chinese: 豆豉), which is probably the culinary ancestor of all fermented soybean products. The earliest archaeological evidence of *douchi* dates back to the early Han dynasty, around 165BC, where *shi* (豉) was found clearly marked in Eastern Han Tomb No. 1, unearthed in 1972 by Chinese archaeologists.[210]

Modern day *douchi*, sometimes referred to in the west as 'fermented black soybeans' or 'soy nuggets', is made by soaking and cooking soybeans, inoculating the beans with either with a pure culture or plant materials, and fermenting the cooked soybeans for several days, to create *douchi qū* (豆豉麹). The fermented soybeans are mixed with salt, sometimes ginger and other spices, brined for several months, then dried.[210] The resulting dried, salty, savoury bean has a complex taste and is used as a seasoning in many different dishes, and to create other seasonings and sauces.

Due to the size and diversity of China, methods of making *douchi* differ considerably, and there is no one correct method. Researchers often differentiate between types of *douchi* based on the microorganism used, which can be fungi of the genera *Mucor*, *Aspergillus* and *Rhizopus*, or bacterial, typically *Bacillus*.[211]

Both *Bacillus subtilis* and *Bacillus amyloliquefaciens* have been isolated from douchi,[212,213] and these two closely related sporulating *Bacillus* species are commonly found in rice straw samples (see Chapter IV.6). The primary difference between these two species is *Bacillus amyloliquefaciens* is much more proficient at breaking down starch into sugar. Working together, a mixed culture of these two bacteria can produce a mixture of proteases and amylases just like fungi-based *douchi qū*.

China: Sui-douchi

Sui-douchi (Chinese: 水豆豉) is commonly found in the southern Chinese provinces of Yunnan, Szechuan, and Guizhou. This version of *douchi* also uses bacteria to ferment cooked yellow soybeans, which is then mixed with salt, peppers, and other ingredients and brined for a less than a week, and served without drying. Unlike normal *douchi*, it is used more as a primary protein source, such as a topping on rice or noodles, rather than a seasoning.

2. Etymologically related Natto in Japan

Japan: Hama natto and Daitokuji natto

In Japan, a few lesser-known food products contain the name 'natto', although they are different from *itohiki natto*. These foods don't meet our criteria for 'natto-like' foods since they are not bacterial ferments, and only share the etymological name of 'natto' and a common culinary origin.

Hama natto (浜納豆), and *daitokuji natto* (大徳寺納豆) closely resemble Chinese *douchi* in taste and function. These two products are sometimes referred to as *shiokara natto* (塩辛納豆), or salted natto, and are made by creating a soybean *koji*, then allowing it to brine for one to several months depending on the method. These salted natto products are typically made in small batches in temples. *Hama-natto* is distinguished into *Yamaya*, *Horonji* and *Daifukuji* types based on its temple of origin.

Itohiki natto was likely developed from the production of *shiokara-natto* when temple monks decided to try the fermented soybeans without proceeding to the brining step.

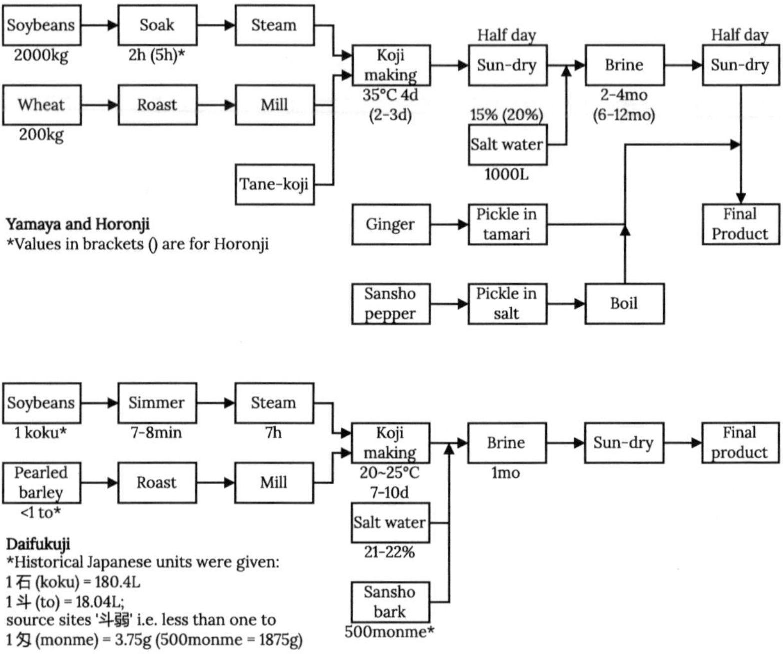

Figure VII.2 Methods for making hama natto.[214,215]

By limiting the water absorbed by the soybeans, and mixing the steamed soybeans with dry, milled wheat or barley, the fermenting soybeans have a considerably lower surface moisture content than *itohiki natto*, which favours the growth of fungi such as *Aspergillus spp.*

The final product has a water content of 29–39%, and a salinity of 9.3%–14.2%, making it compositionally similar to a soybean miso. It's often eaten with rice porridge, as a topping for *ochazuke*, among others. The taste is like pure-soybean and darker misos such as *hacho-miso,* which are generally much more popular than *hama-natto* or *daitokuji-natto*.

No Natto in the Brewery!

Bacillus subtilis is considered one of the greatest threats by Japanese brewers of *sake, shochu, shoyu* and *miso* trying to create *koji* for their respective brews. Slimy *koji* is a dead giveaway that a batch of *koji* has been ruined by natto bacteria, and brewers will often ask visitors and employees not to eat natto when visiting or working in a brewery. It's unclear whether this is just a superstition, and no formal scientific studies have been completed on the matter.

Japan: Ama Natto

In Japan, there is a product called *ama natto* (甘納豆) which is simply a confectionary made by cooking beans in sugar. This product only shares the etymological name of 'natto'.

3. Natto-like Foods in the Himalayas, Southeast Asia

Travelling south, many of the countries in the Himalayas and Southeast Asia have fermented soybean products similar to natto.

A few common themes emerge among these products. Firstly, they are usually made in village and household settings, and use atmospheric boiling rather than pressure steaming to cook the beans until soft. Secondly, they all use naturally occurring bacteria from any of the plant materials that come into contact with the cooked soybeans during fermentation. Most of them are fried, dried, or cooked to make curries and side dishes. All of these preparations require the washing and soaking the beans prior to cooking, so these steps will not be described in detail.

Thailand: Thua Nao

In Northern Thailand (the area including Chiang Mai, Chiang Rai, Lampang, and Mae Hong Son), we find the traditional soy food *thua nao* (Thai: ถั่วเน่า).

The steps to make *thua nao* are as follows: soybeans are washed thoroughly and soaked overnight, boiled for 3–4 hours until soft, then placed in bamboo baskets lined with banana leaves to undergo naturally inoculated fermentation for 2–3 days. The resulting ferment has an ammonia odour and a slight stickiness.[216]

This can be consumed directly, but the more common application is to cook it by steaming or roasting, sometimes with additional ingredients such as chilis, onions, and salt. It is also common to flatten out cooked *thua nao* into flat disks to be sun-dried into *thua nao kab*.[216]

The banana leaves are a source of epiphytic *Bacillus* species, primarily *B. subtilis*, but researchers have also isolated *B. licheniformis*, *B. megaterium*, and *B. pumilus* from *thua nao*.

Nepal: *Kinema*

In Nepal, *kinema* (Nepalese: किनेमा) is made from small sized yellow soybeans, or local dark brown soybeans.

The steps to make *kinema* are as follows: soybeans are washed thoroughly and soaked overnight, immersed in fresh water, and boiled for 2–3 hours until soft. Boiled beans are split roughly in half using a wooden mortar and pestle, then about 1% by weight of firewood ash is added to the cooked, split soybeans. This mixture is fermented in a bamboo basket lined with fresh ferns, fig leaves, and banana leaves. The basket is covered and left to ferment at 25–40°C, sometimes above an earthenware oven, for 1–3 days depending on the time of year.[217]

Kinema is typically fried directly in oil to make a curry, but it can also be dried, then cooked with leafy vegetables to make a curry.

Once again, *B. subtilis* is the primary microorganism behind the fermentation of *Kinema*. Other species of *Bacillus* have also been isolated, including *B. licheniformis*, *B. cereus*, *B. circulans*, *B. thuringiensis*, and *B. sphaericus*, as well as some lactic acid bacteria and yeasts. These microorganisms can be introduced at any step of the process, since none of the tools used, such as the mortar and pestle, bamboo basket, and leaves, are sanitized. In academia, there have been developments to use pure cultures of *B. subtilis* prepared in the laboratory as starter cultures.[217]

Cambodia, Laos: *Sieng*

In Cambodia and Laos, *sieng* (Khmer: សៀង) is prepared by spreading boiled soybeans on bamboo baskets and allowing the beans to naturally ferment for 2 days. They are then soaked in saltwater for 5–7 days immediately after, and sometimes tree sap or enzymes are added to this brine. *Sieng* is used as a seasoning with salt and spices.[217] The brief brining period makes it similar to Chinese *sui-douchi*, and Laos directly borders the province of Yunnan, so the two products may have a close origin.

Myanmar: *Pepok*

In Myanmar, *pepok* is prepared by wrapping boiling soybean in leaves, then leaving them to naturally ferment for 2–4 days. These fermented beans are mashed with salt and hot pepper, and sun-dried to be used as a seasoning, or roasted.[217] This makes it similar to Thai *thua nao.*

4. Natto-like Foods in Northern India

India is such a vast country with multiple dialects and cultures that it warrants its own section to describe each of the natto-like foods, which are primarily found in the northeastern, Himalayan regions.

Manipur, India: *Hawaijar*

In Manipur, *hawaijar* is prepared by packing boiled soybeans into lidded, bamboo baskets lined with fig leaves or banana leaves, then loosely covered, and left to ferment for 3–5 days. It may be used as-is, or sun dried for 2–3 days. *Hawaijar* is eaten directly as a condiment or mixed with vegetables to make a curry.[217]

Meghalaya, India: *Tungrymbai*

In Meghalaya, *tungrymbai* (Hindi: तुंगरिमबाई) is prepared by packing boiled soybeans into bamboo baskets with the leaves of 'Cool Mat', or *Schumannianthus dichotomus*, then loosely covered and left to ferment for 3–5 days. Notably, the soaked soybeans are hulled before cooking, and during the cooking process, all the cooking liquid is reduced or fully absorbed. The resultant soybean ferment is mashed with water and cooked while stirring continuously, until the water has reduced, then the mash is mixed with other ingredients to make a curry.[217]

Nagaland, India: *Aakhone*

In Nagaland, *aakhone* (Nagamese: আখুনী) is made by wrapping cooked soybeans in fresh banana leaves, or similar leaves of *Phrynium pubinerve* or

Macaranga indica, then left to ferment above a fireplace for 5–7 days. The fermented beans can be made into cakes, then dried in an oven, or separated into individual beans for 2–3 days of sun drying. *Aakhone* can be made into pickles by mixing with vegetables or cooked with other proteins as a dish.[217]

Mizoram, India: *Bekang*

In Mizoram, *bekang* is made by wrapping cooked soybeans in fresh leaves of *Calliparpa aroria* or *Phrynium spp.*, packing them in a bamboo basket, then allowing them to ferment naturally for 3–4 days. *Bekang* is eaten as is, or made into a curry.[217]

Arunachal Pradesh, India: *Peruyaan*

In Arunachal Pradesh, *peruyaan* is made by packing cooked soybeans in a bamboo basket lined with fresh ginger leaves, then left to ferment above a fireplace for 3–5 days. *Peruyaan* is consumed as a side dish after mixing with hot water, chilis, and salt, without further cooking.[217]

5. Trends Across Natto-like Foods

5.1 Applications and Origins

Interestingly, Japanese natto seems to be the only product out of the 'natto-like foods' that is commonly consumed as-is without any further processing. The other products find their way into stews, curries, and brines, sometimes with additional cooking or drying steps. Perhaps these steps are necessary for the destruction of potentially pathogenic bacteria that can occur in naturally inoculated ferments.

Climate may also play a factor in the consumption of non-cooked natto. Of all the abovementioned geographic regions, only high altitude areas of the Himalayas, and the northern areas of Korea, Japan, and China experience cold winters, during which the shelf life of fermented soybeans can be extended for later consumption. Otherwise, drying and cooking would be preferrable to make a shelf-stable product. We know from Chapter I.3 that natto made during the Edo period was primarily peddled during winter for this reason.

The Natto Triangle and 'KNT Triangle'

Many of the initial findings on natto-like foods in East and Southeast Asia were documented by Japanese researchers, who travelled around soybean producing nations during the postwar period to document these ferments. Japanese ethnologist Sasuke Nakao first proposed the idea of a 'Natto Triangle' in 1972. This triangle, when plotted on a map, has vertices located in Japan (natto), Nepal (*kinema*) and Indonesia (*tempeh*).[1] Within this triangle, there is an abundance of natto-like foods.

Later in 2010, Dr. Jyoti Tamang of Sikkim, India proposed a revision to this triangle concept called the KNT Triangle. It has the same vertices except the southeastern vertex ends in Thailand's *thua nao*, rather than Indonesia's *tempeh*, since the latter is not bacterial ferment.[218]

Even later in 2019, Korean researchers diving deep into the origins of *cheonggukjang* proposed that this geographic concept should be called a *cheonggukjang belt,* after providing a historical argument that all these foods are tied to the Korean fermented soybean product.

To argue for one triangular shape, a belt, or otherwise seems to be as fruitless as arguing over the name of a bacterium. There's probably a bit of bias involved depending on the nationality of the researcher. However, in terms of market size and per-capita consumption, and the scale of research, technology, and industrialization, natto and *douchi* appear to dwarf the other ferments.

Natto-like Ferments, Buddhism and Veganism

Putting aside geographical shapes, all these foods find themselves intertwined with Buddhism, although this might also be happenstance since the spread of religion may occur simultaneously with the spread of agricultural technology. The practice of vegetarianism and veganism in Buddhism varies by location and school of thought, but it's likely that the techniques associated with fermenting soybeans were developed by Buddhist monks trying to improve the taste of boiled soybeans, which are a valuable source of plant-based protein.

When looking at the production process of natto-like products, it seems almost obvious that a bacterial ferment would be discovered accidentally. It's as simple as: boiling soaked beans, storing them in a container with plant matter (which will contain epiphytic *Bacillus spp.* or its endospores), and letting it sit somewhere warm and moist, such as a tropical environment or near a fire.

5.2 Bacterial Species

Bacillus subtilis is expectedly the dominant species found in all the sampled natto-like products. Other species of *Bacillus* that have been isolated include *B. amyloliquefaciens*, *B. megaterium*, *B. licheniformis*, and *B. pumilus*. These species are all aerobic or facultative aerobic spore-forming bacteria, making them ubiquitous on dried plant matter and well-suited to colonizing hot soybeans. *B. amyloliquefaciens* might help the other bacteria by liberating more free sugars from soybean starch. *B. megaterium* is a producer of Vitamin B_{12}, which may be important since vegan versions of Buddhist cuisine can lack a source of this vitamin.

There are several reasons why *Bacillus* is favoured over fungi in these ferments. Firstly, in the battle between bacteria and fungi on the surface of a cooked soybean, it comes down to small differences in intrinsic and extrinsic factors that favour one kingdom of life over the other in a spontaneous fermentation. Bacteria prefer warmer temperatures and higher moisture environments compared to fungi, and different species of each kingdom have different preferred pH ranges.

For example, in the fermentation of the soybean *koji*, a Japanese fungal ferment using *Aspergillus spp.*, water absorption during the soaking step is limited to 1.6x the original soybean's weight, so that the steamed soybeans have a lower moisture content that favours fungi. Soybean koji is also mixed with dried, toasted wheat or barley flours. By comparison, natto uses fully soaked beans at 2.2–2.3x the original soybean's weight.

Secondly, consider the Indonesian product *tempeh*, which is made by wrapping cooked soybeans in banana leaves. We know that original preparations of *tempeh* rely on the souring of soybeans during the soaking step by lactic acid bacteria. This would easily occur in the tropical heat of Indonesia. Recall that lactic acid is detrimental to the growth of *Bacillus subtilis* (see Chapter II.1.1) and occurs in high temperature soaking water. The resultant lactic acid and drop in pH favours the growth of *Rhizopus spp.*, the fungi responsible for tempeh. Once cultures of fungal spores were developed, *tempeh* fermentation would always proceed with pure *Rhizopus* cultures, and sometimes vinegar is added to lower the pH.

5.3 Fibrinolytic Proteases and Bioactive Compounds

Detailed biochemical analysis revealed that many wild strains of *Bacillus* used in natto-like foods were found to produce proteases that degrade fibrin,[212,213,219] so the fact that Japan's natto-producing *Bacillus subtilis* produces a fibrin-degrading subtilisin is happenstance, but not a particularly rare feature.

However, even for natto-like foods that contain fibrinolytic enzymes, these enzymes must present in a sufficiently large serving, be uncooked to preserve its enzymatic activity, and survive the acidity of the stomach before it can be absorbed as a potential bioactive substance for lowering blood pressure. These natto-like foods face the same challenges as nattokinase described in Chapter VI.2.2 if they are to be considered a possible treatment to hypertension.

5.4 Future of Natto-like Foods and Natto Overseas

It's unclear if the natto-like foods of the Himalayas, India, and Southeast Asia will undergo mass industrialization like natto. This will depend on their future popularity, and the incorporation of companies to mass produce the product.

Chinese *douchi* is already ubiquitous and produced on an industrial scale by many manufacturers in China. The same applies to *cheonggukjang*, although in Korea, natto is also easy to obtain. Korean food companies manufacture natto domestically, indicating that Korean demand is large enough to support natto factories, rather than relying on imported Japanese natto.

Market data on the import of Japanese natto is not widely available. Anecdotally, frozen Japanese and Korean natto are easy to find in Asian supermarkets overseas. Natto producers in the west remain relatively small-scale and usually consist of small-batch processors selling to specialty health food stores. Although the natto-making process is relatively simple, a large market demand is still needed to invest in large-scale natto making equipment.

Summary

When examining the entire family of natto-like foods across Asia, it becomes clear that bacterial-fermented soybeans are common everywhere that soybeans were historically grown. The methods to ferment these soybeans may have evolved naturally from storing cooked soybeans in a warm place with plant materials, which are a bountiful source of *Bacillus* endospores. The techniques may also have spread alongside soybean agriculture and Buddhist cuisine.

Interestingly, natto is the only fermented soybean product that is most commonly consumed in its raw form, rather than as a cooked, dried, or otherwise processed ingredient in other dishes.

Further Reading

For a detailed examination of fermented soy foods across Asia, I highly recommend any of the works by Dr. Jyoti Prakash Tamang, who has authored, edited, or coauthored multiple books and papers on the subject:

- **Naturally fermented ethnic foods of India**[220]
- **Health Benefits of Fermented Foods and Beverages**[221]
- **Himalayan Fermented Foods: Microbiology, Nutrition, and Ethnic Values**[218]
- **Fermented Foods and Beverages of the World**[222]

The Future of Natto

I think natto has a bright future overseas along with a steady market in Japan. Markets everywhere have demonstrated an increasing openness to new and foreign foods, and a positive attitude towards environmental consciousness and the consumption of plant-based proteins, of which soybeans are one of the most productive and nutritious sources. In this last section, I want to write down my thoughts on the future of natto, after learning as much as possible through writing the contents of this book.

Developments in Natto Production Technology

Japanese natto technology is already extremely well-optimized, and both large and small scale natto factories are highly efficient. This technology will likely be exported if natto gains more popularity overseas. It's possible to design natto-making equipment from scratch, but it would be prudent to use Japanese equipment, either imported or as a reference design. Even though natto-making is a relatively simple process, it would be foolhardy to disregard nearly a century of technological expertise. In manufacturing, it's this experience—which includes minute details like the layout of equipment, the size of ventilation ductwork, maintenance schedules, and manufacturing best practice—that allows Japanese manufacturers to sell natto at such a low cost.

On the other hand, single-use plastics seem to be falling out of favour, so I expect the iconic square, polystyrene paper (PSP) containers commonly used in natto to be gradually replaced with alternatives. Luckily, paper cups are already the second most common container type, and other container types and biodegradable plastic films like polylactic acid might see widespread adoption.

Marketing and Natto Products

The biggest challenge to the internationalization of natto might simply be its 'unique' qualities. Natto's appearance, taste, texture, and aroma are challenging to say the least, especially for anyone who didn't grow up eating it. Despite this, the international market for natto seems to be growing, especially among adventurous eaters, and environmental and health-conscious consumers.

It's not clear whether non-soybean types of natto will become popular. Creating natto out of other beans is possible, but strains of common beans have not been bred specifically to make natto. Non-soybean natto would require some research and development on both the agricultural and processing ends. The fact that fermentation of beans with *B. subtilis* removes flatulence-causing sugars is interesting, and worth looking into. I wonder why 'enzyme-treated' beans are not a thing. To be frank, however, common beans taste just fine, and don't suffer the same culinary drawbacks as cooked soybeans.

Meanwhile in Japan, there is plenty of diversity in natto products. Small natto producers have embraced many strains of domestic soybeans, include ones with strikingly different colours, all with unique tastes and textures. I hope these soybean strains will eventually find their way into the hands of growers and producers overseas. Large and small natto producers also already offer sauces and flavours beyond the original *tare* and *karashi*, and package natto with a huge range of sauce options.

Natto Bacteria

Research into *Bacillus subtilis* will continue, and I hope new strains are developed to produce different types of natto. Although public opinion on genetically modified organisms is unfavourable, I think these tools should be explored, especially if they can improve the taste and functionality of natto, and the safety behind these modifications are well-understood. Exploring and screening wild strains of *Bacillus subtilis* should also continue, in order to develop new and novel production strains.

Coculturing *Bacillus subtilis* with other species like *Bacillus amyloliquefaciens* (for its ability to break down starch) and *Bacillus megaterium* (for its ability to produce Vitamin B12) should be investigated. These strains have already been isolated from natto-like products, and we know that the original samples of bacteria isolated from pre-modern straw natto were mixed cultures before the development of pure *natto-kin*. Sometimes I wonder if pre-modern natto tasted better or different than modern scientific natto.

If we disregard *B. subtilis* for a moment, there may be an opportunity to develop an entirely novel 'natto' made with some other microorganism that has stronger nutritional, physiological, or probiotic benefits. After all, the creation of natto is merely a coincidence that spore-forming, epiphytic strains of *Bacillus* are common on rice straw and other agricultural materials. This shouldn't exclude researchers from testing soybean fermentation with the massive collection of microorganisms and techniques available to science.

Nutrition and Physiology

The primarily health benefits of natto stem from the fact that you are eating a whole grain soybean product that is a source of Vitamin K_2. Additional benefits from some of the functional compounds, including nattokinase and isoflavones, are somewhat poorly understood. Some of these physiological benefits of natto need further research, but at least they are promising. Research into these topics, including studies of biochemical mechanisms, large scale randomized controlled trials, and epidemiological studies, will continue.

Future Editions of this Book

Finally, I plan on maintaining future editions of this book. By the time the next edition is released, hopefully some of the questions and issues posed above will have been addressed or researched. As stated before, science is constantly evolving, and our understanding of every topic in this book is bound to develop with time.

In the future, I'd like to include more details about the natto factory, including in-depth process details, machine designs, and more photos. I'd also like to do a detailed analysis of all natto product offerings, including those of the smaller overseas natto companies, and take a look at the challenges and innovations required to produce natto outside of Japan's well-established natto infrastructure.

Keith W.

Glossary of Japanese Terms

Japanese	Romaji	English, Definition and Notes
Natto Etymology		
納豆	nattō	
納所	nassho	Temple kitchen, workshop, or the monk who works in this location
大豆	daizu	Soybean
豉	kuki shi	Pre-cursor to natto, similar to douchi; historical term
塩辛納豆	shiokara nattō	Salted natto
糸引き納豆	itohiki nattō	Threaded-forming (sticky) natto
叩き納豆	tataki nattō	Chopped natto
納豆汁	nattō jiru	Natto soup
納豆売り	nattō uri	Natto peddler
浜納豆	hama nattō	Soldier's natto (type of salted natto)
大徳寺納豆	daitokuji nattō	Daitokuji temple natto (salted)
甘納豆	ama nattō	Sweet natto (confectionary)
Grain Size and Package Label		
粒納豆(つぶ納豆)	tsubu nattō	Whole grain natto
丸大豆納豆	maru daizu nattō	Whole-grain soybean natto, synonymous with above
挽き割り納豆 (ひきわり納豆)	hikiwari nattō	Crushed natto
極大粒	goku ōtsubu	Extra large grain (not an official size)
大粒	ōtsubu	Large grain
中粒	nakatsubu	Medium grain
小粒	kotsubu	Small grain
極小粒	goku kotsubu	Extra small grain
国産	kokusan	Domestic (Japanese)
有機	yūki	Organic
たれ	tare	Sauce
からし	karashi	Mustard
たれ無し(たれなし)	tare nashi	No sauce

Japanese	Romaji	English, Definition and Notes
たれ付	tare tsuki	With sauce
金のつぶ	kin no tsubu	'Golden grains' (Mizkan's brand for low odour natto)
豆乃香(豆の香)	mamenoka	'Smell of natto' (Kusanago foods' brand for non-sticky natto)
Companies & Associations		
宮城野納豆製造所	miyagino nattō seizōjo	Miyagino Natto Factory Co., Ltd.
高橋祐蔵研究所	takahashi yūzō kenkyūjo	Yuzo Takahashi Laboratory Co., Ltd.
成瀬醗酵化学研究所	naruse hakkō kagaku kenkyūjo	Naruse Fermentation Chemical Laboratory Co., Ltd.
全国納豆協同組合連合会	zenkoku nattō kyōdō kumiai rengō-kai	Japan Natto Manufacturers Cooperative Society Federation (JNCSF) aka 'Natto-Ren'
納豆組合	nattō kumiai	Natto cooperative (1939 historical association, prefecture level)
全国納豆工業組合協会	zenkoku nattō kōgyō kumiai kyōkai	Japan natto industrial cooperative association (1941 historical association) (JNICA)
鈴与工業株式会社	suzuyo kōgyō kabushikigaisha	Suzuyo Industry Co., Ltd. (Suzuyo Kogyo)
みわ精機株式会社	miwa seiki kabushikigaisha	Miwa Machinery Co., Ltd. (Miwa Seiki)
Raw Material Handling		
俵	tawara (as a noun) hyō (as a unit)	'Bale'-historical volumetric unit of a rice straw bag; approximately equivalent to 60kg soybeans, dry
原料大豆	genryō daizu	Raw material soybeans
貯蔵	chozō	Storage
投入ホッパー	tōnyū hoppā	Input hopper
昇降機	shōkōki	Elevator
サイロタンク	sairo tanku	Silo tank
コンベアー	konbeā	Conveyor
昇降機	shōkōki	Elevator
粗選	arasen	Rough selection
粗選機	arasenki	Rough selection machine
石抜	ishinuki	Stone removal
石抜機	ishinukiki	Stone removal machine

Japanese	Romaji	English, Definition and Notes
研磨	kenma	Polishing
研磨機	kenmaki	Polishing machine
粒形選別	tsubugata senbetsu	Grain shape shorting
粒形選別機	tsubugata senbetsuki	Grain shape sorter
色彩選別	shikisai senbetsu	Color selection
色彩選別機	shikisai senbetsuki	Color selection machine
金属除去	kinzoku jokyo	Metal removal
金属選別機	kinzoku senbetsuki	Metal detection machine
精選大豆	seisen daizu	Processed soybeans
計量	keiryō	Metering
計量器	keiryōki	Metering (machine for dosing soybeans from storage to washing)
Washing, Soaking, Steaming, Inoculation		
洗豆	arai mame	Washing beans
豆洗機	mame araiki	Bean washing machine
洗浄	senjō	Washing
洗浄機	senjōki	Washing machine
水分離器	mizu bunriki	Water separator (vortex type)
浸漬	shinseki	Soaking
浸漬タンク	shinseki tanku	Soaking tank
蒸煮	jōsha	Steaming
蒸煮缶	jōsha kan	Steaming can (steamer)
煮豆	nimame	Steamed beans
煮豆台車	nimame daisha	Steamed beans cart (i.e. to unload steamer and move to filling machine)
接種	sesshu	Inoculation
滅菌水	mekkinsui	Sterilized water
納豆菌	nattō kin	Natto starter (endospores)
希釈菌液	kishaku kineki	Diluted bacterial solution
容器	yōki	Container-natto packaging (for natto)
充填	jūten	Filling (primary)
盛込み充填機	morikomi jūtenki	Filling machine

Japanese	Romaji	English, Definition and Notes
フィルム	firumu	Film (i.e. atop natto)
重量チェック	jūryō chekku	Weight check (scale to verify filling)
金属チェック	kinzoku chekku	Metal check (metal detector)
手盛り機	temori ki	Hand filling machine
オートケーサー	ōtokēsā	'Autocaser' (loading into containers)
インケーサー	inkēsā	'In-caser' (loading into containers)
箱入機	hakoiri ki	Boxing machine (loading into containers)
コンテナ	kontena	Container (i.e. stackable plastic trays, not to be confused with individual containers for natto/steamed beans)
コンテナ詰込み	kontena tsumekomi	Container filling
コンテナ台車	kontena daisha	Container cart (i.e. dolly for moving stacked containers)
Packaging Materials		
ポリステレン	porisuteren	Polystyrene (for PSP packaging)
紙カップ	kami kappu	Paper cups
プラスカップ	purasu kappu	Plastic cups
塩ビカップ	shiobi kappu	PVC cups
ポリ袋	pori bukuro	Polyethylene bag (PE lined kyogi)
経木	kyōgi	Paper thin sheet of wood
すだれ	sudare	Bamboo screen
苞（つと）	tsuto	Straw bundle (typically rice straw)
藁（わら）	wara	Straw (typically rice straw)
Fermentation		
発酵	hakkō	Fermentation (technically respiration/incubation)
発酵室	hakkōshitsu	Fermentation room
熟成	jukusei	Aging (post fermentation ripening)
冷蔵	reizō	Refrigeration
一次冷蔵	ichiji reizō	Primary refrigeration
温度	ondo	Temperature

Japanese	Romaji	English, Definition and Notes
湿度	shitsudo	Humidity
時間	jikan	Time
品温	hinon	Product temperature
室温	shitsuon	Room temperature
予冷	yorei	Precooling
発芽繁殖	hatsuga hanshoku	Germination and reproduction
誘導期	yūdōki	Induction period
対数期	taisūki	Exponential growth
定常期	teijōki	Stationary phase
死滅期	shimetsuki	Death phase
Secondary Packaging and Distribution		
コンテナ箱出し	kontena bakodashi	Container unloading (removal of natto packages from container trays)
コンテナ洗浄	kontena senjō	Container cleaning
コンテナ乾燥	kontena kansō	Container drying
アンケーサー	ankēsā	'Uncaser' (loading out of containers)
取り出し機	toridashi ki	Unloading machine (loading out of containers)
包装	hōsō	Packaging (secondary)
各種包装機	kakushu hōsōki	Various packaging machines
段ボールケーシング	danbōru kēshingu	Cardboard casing
段ボールケーサー	danbōru kēsā	Cardboard caser
日付	hizuke	Date (of distribution or expiry)
製品冷蔵	seihin reizō	Product refrigeration
仕分	shiwake	Sorting
出荷	shukka	Shipping
出荷用冷蔵庫	shukkayō reizōko	Refrigerator for shipping (secondary refrigeration)
配送	haisō	Delivery
Quality Evaluation		
菌の被り	kin no kaburi	Coverage of bacteria
溶菌状態	yōkin jōtai	Bacteriolytic condition

Japanese	Romaji	English, Definition and Notes
豆の割れ，つぶれ	mame no ware, tsubure	Cracked or crushed beans
豆の色	mame no iro	Colour of beans
香り	kaori	Smell
硬さ	katasa	Hardness
味	aji	Taste
糸引き	itohiki	Stickiness
総合評価	sōgō hyōka	General
Products made from Natto		
干し納豆	Hoshi nattō	Dried natto (typically air dried)
ドライ納豆	Dorai nattō	Dried natto (typically oil fried)
フリーズドライ納豆	Furizu dorai nattō	Freeze dried natto
雪割り納豆	Yukiwari nattō	Natto preserved with koji
ゴト納豆	Goto nattō	Synonymous with above
とうぞう	Touzou	Natto preserved with daikon, soybean broth, others
そぼろ納豆	Soboro nattō	Natto preserved with daikon, others
Selected Historical Figures (Family Name – Given Name) and Texts		
矢部規矩治	Yabe Kikuji	First to analyze natto scientifically
澤村真 (澤村博士)	Sawamura Shin (Sawamura Hakase*)	Isolated natto producing strain of *B. subtilis* (*Professor Sawamura)
村松舜祐	Muramatsu Shunsuke	Isolated natto producing strain of *B. subtilis*
成瀬金太郎	Naruse Kintaro	Modern natto process; spore producer
半澤洵	Hanzawa Jun	Modern natto process
三浦二郎	Miura Jiro	Modern natto process; spore producer
高橋三雄次	Takahashi Mitsuji	Spore producer
須見洋行	Sumi Hiroyuki	Discovered nattokinase
渡辺杉夫	Watanabe Sugio	Prolific modern natto author
石井泰二	Ishii Taiji	Prolific natto blogger

References

(1) Shurtleff, W.; Aoyagi, A. *History of Natto and Its Relatives (1405-2012): Extensively Annotated Bibliography and Sourcebook*; Soyinfo Center: Lafayette, CA, 2012.

(2) Walker, B. L. *A Concise History of Japan*; Cambridge concise histories; Cambridge University Press: Cambridge, 2015.

(3) Yano, K.; Kaner, S. Early Agriculture in Japan. In *The Cambridge world history. Volume II, A world with agriculture, 12,000 BCE - 500 CE*; Cambridge Histories Online; Cambridge University Press: Cambridge, 2015; pp 353–386.

(4) Sansom, G. B. *A History of Japan to 1334*; Stanford studies in the civilizations of eastern Asia; Stanford University Press: Stanford, Calif., 1958.

(5) Kwon, D. Y.; Chung, K. R.; Jang, D.-J. The History and Science of Chongkukjang, a Korean Fermented Soybean Product. *Journal of Ethnic Foods* **2019**, *6* (1), 5. https://doi.org/10.1186/s42779-019-0004-8.

(6) Follador, E. Re-Contextualizing Shōjin Gyorui Monogatari, or When the Fish Declared War on the Greens. International Research Center for Japanese Studies March 2021. https://doi.org/10.15055/00007601.

(7) Opal, S. M. A Brief History of Microbiology and Immunology. *Vaccines: A Biography* **2009**, 31–56. https://doi.org/10.1007/978-1-4419-1108-7_3.

(8) Kikuji, Y. On the Vegetable Cheese, Natto. *Bull. Coll. Agric., Tokyo Imp. Univ.* **1894**, *2* (2), 68–72.

(9) Sawamura, S. On the Micro-Organisms of Natto. *Bull. Coll. Agric., Tokyo Imp. Univ.* **1906**, 7 (1), 107–110.

(10) Shunsuke Muramatsu. On the Preparation of Natto. *Journal of the College of Agriculture. Imp Univ Tokyo 5* (1), 81–94.

(11) Sawamura, S. On Bacillus Natto. *J. of the College of Agriculture, Tokyo Imp. Univ.* **1913**, *5* (2), 189–191.

(12) 納豆の発生と歴史 [Origin and History of Natto]. In *納豆入門 [Introduction to Natto]*; 日本食糧新聞社 [Nihon shokuryō shinbunsha]: Tokyo, 2020; pp 1–14.

(13) 国元堀田; 博佐々木. 近代納豆の幕開けと応用菌学 [The Dawn of Modern Natto and Applied Mycological Science]. 化学と生物 **2011**, *49* (1), 57–62. https://doi.org/10.1271/kagakutoseibutsu.49.57.

(14) 近代納豆工業への展開 [Expansion into modern natto industry]. In *納豆入門 [Introduction to Natto]*; 日本食糧新聞社 [Nihon shokuryō shinbunsha]: Tokyo, 2020; pp 15–28.

(15) Jun Hanzawa. *納豆製造法 [Natto Production Method]*, 2nd ed.; 札幌納豆容器改良会 [Sapporo Nattō Yōki Kairyōkai]: Sapporo, 1930.

(16) Shunsuke Muramatsu; Kinataro Naruse. *最新納豆製造法 [The Latest Natto Manufacturing Method]*, 1st ed.; 明文堂 [Meibundo], 1950.

(17) *成瀬醗酵化学研究所 [Naruse Fermentation Chemistry Laboratory]*. 納豆 wiki. https://seesaawiki.jp/w/taiji141/ (accessed 2022-12-04).

(18) *Yuzo Takahashi Research Institute - Information*. https://web.archive.org/web/20080202024655/http://www.nattomoto.net/annai.html (accessed 2022-12-04).

(19) *Japan Natto Cooperative Society Federation - Overview*. http://natto.or.jp/about/index.html (accessed 2022-12-04).

(20) Kiuchi, K.; Watanabe, S. Industrialization of Japanese Natto. In *Industrialization of Indigenous Fermented Foods*; Food science and technology (Marcel Dekker, Inc.); Marcel Dekker: New York, 2004.

(21) 「納豆」を知って原料大豆を選ぶ [Knowing "Natto" and selecting soybeans]. In *納豆: 原料大豆の選び方から販売戦略まで [Natto: From how to select raw soybeans to sales strategies]*; 農山漁村文化協会 [Nō-san-gyoson Bunka Kyōkai]: Tokyo, 2002; pp 23–55.

(22) 現代の納豆製造、流通・保蔵、および表示 [Modern natto production, distribution, storage, and labeling]. In *納豆入門 [Introduction to Natto]*; 日本食糧新聞社 [Nihon shokuryō shinbunsha]: Tokyo, 2020; pp 29–51.

(23) Woerfel, J. B. Harvest, Storage, Handling, and Trading of Soybeans. In *Practical Handbook of Soybean Processing and Utilization*; Erickson, D. R., Ed.; AOCS Press, 1995; pp 39–55. https://doi.org/10.1016/B978-0-935315-63-9.50008-5.

(24) Spencer, M. R. Effect of Shipping on Quality of Seeds, Meals, Fats, and Oils. *Journal of the American Oil Chemists' Society* **1976**, *53* (6, Part 1), 238–240. https://doi.org/10.1007/BF02605691.

(25) 品質のよい納豆を安定して生産する [Stable production of high quality natto]. In *納豆 : 原料大豆の選び方から販売戦略まで [Natto: From how to select raw soybeans to sales strategies]*; 農山漁村文化協会 [Nō-san-gyoson Bunka Kyōkai]: Tokyo, 2002; pp 57–83.

(26) 納豆原料 [Natto Raw Materials]. In *納豆の科学 [Advanced Science on Natto]*; Kiuchi, K., Nagai, T., Kimura, K., Eds.; 建帛社 [Kenpakusha]: Tokyo, 2008; pp 1–30.

(27) 納豆製造技術の基礎知識 [Basic knowledge of natto production technology]. In *納豆入門 [Introduction to Natto]*; 日本食糧新聞社 [Nihon shokuryō shinbunsha]: Tokyo, 2020; pp 52–83.

(28) *Canadian Soybeans – Food Grade: Soy Canada*. https://soycanada.ca/buy-canadian-soybeans-food-grade/ (accessed 2022-12-07).

(29) Government of Canada, C. G. C. *Quality of Canadian food-type soybeans 2020*. https://www.grainscanada.gc.ca/en/grain-research/export-quality/pulses/soybean-food/2020/07-natto.html (accessed 2022-05-30).

(30) 納豆の製造・流通・保蔵 [Natto Production, Distribution and Storage]. In *納豆の科学 [Advanced Science on Natto]*; Kiuchi, K., Nagai, T., Kimura, K., Eds.; 建帛社 [Kenpakusha]: Tokyo, 2008; pp 31–42.

(31) 納豆製造法 [Natto Production Method]. In *納豆の研究法 [Investigative Approach to Natto]*; Kan, K., Nagai, T., Kimura, K., Kodaka, K., Muramatsu, K., Watanabe, S., Eds.; 恒星社厚生閣 [Kōseishakō seikaku]: Tokyo, 2010; pp 47–60.

(32) 挽き割り納豆および加工納豆類 [Hikiwari natto and processed natto]. In *納豆入門 [Introduction to Natto]*; 日本食糧新聞社 [Nihon shokuryō shinbunsha]: Tokyo, 2020; pp 133–141.

(33) Krawczyk, A. O.; de Jong, A.; Omony, J.; Holsappel, S.; Wells-Bennik, M. H. J.; Kuipers, O. P.; Eijlander, R. T. Spore Heat Activation Requirements and Germination Responses Correlate with Sequences of Germinant Receptors and with the Presence of a Specific SpoVA2mob Operon in Foodborne Strains of Bacillus Subtilis. *Appl Environ Microbiol* **2017**, *83* (7), e03122-16. https://doi.org/10.1128/AEM.03122-16.

(34) Muramatsu, K.; Katsumata, R.; Watanabe, S.; Tanaka, T.; Kiuchi, K. Improvement of Itohiki-Natto-Manufacturing Process. *Nippon Shokuhin Kagaku Kogaku Kaishi* **2001**, *48* (4), 277–286. https://doi.org/10.3136/nskkk.48.277.

(35) 納豆製造工程の管理 [Management of Natto Production Process]. In *納豆入門 [Introduction to Natto]*; 日本食糧新聞社 [Nihon shokuryō shinbunsha]: Tokyo, 2020; pp 84–104.

(36) *Pulmuone's natto plant isn't afraid to get sticky*. https://koreajoongangdaily.joins.com/2018/08/02/industry/Pulmuones-natto-plant-isnt-afraid-to-get-sticky/3051426.html (accessed 2023-03-22).

(37) 品質管理 [Quality Control]. In *納豆の研究法 [Investigative Approach to Natto]*; Kan, K., Nagai, T., Kimura, K., Kodaka, K., Muramatsu, K., Watanabe, S., Eds.; 恒星社厚生閣 [Kōseishakō seikaku]: Tokyo, 2010; pp 61–132.

(38) 品質管理 [Quality Control]. In *納豆の科学 [Advanced Science on Natto]*; Kiuchi, K., Nagai, T., Kimura, K., Eds.; 建帛社 [Kenpakusha]: Tokyo, 2008; pp 159–202.

(39) 納豆の生産計画から販売まで [From production planning to sales of natto]. In *納豆: 原料大豆の選び方から販売戦略まで [Natto: From how to select raw soybeans to sales strategies]*; 農山漁村文化協会 [Nō-san-gyoson Bunka Kyōkai]: Tokyo, 2002; pp 86–120.

(40) Suzuyo Kogyo Co., Ltd. *Suzuyo's Fortes*. https://suzuyokogyo.com/english/strength/ (accessed 2022-07-27).

(41) Sanitation in the Food Industry. In *Principles of Food Sanitation*; Food Science Text Series; Springer International Publishing: Cham, 2018; pp 1–18. https://doi.org/10.1007/978-3-319-67166-6.

(42) The Relationship of Microorganisms to Sanitation. In *Principles of Food Sanitation*; Food Science Text Series; Springer International Publishing: Cham, 2018; pp 33–72. https://doi.org/10.1007/978-3-319-67166-6.

(43) Food Contamination Sources. In *Principles of Food Sanitation*; Food Science Text Series; Springer International Publishing: Cham, 2018; pp 83–92. https://doi.org/10.1007/978-3-319-67166-6.

(44) 全国納豆協同組合連合会 [Japan Natto Cooperative Society Federation]. 納豆連 HACCP [Natto-Ren HACCP], 2018. https://haccp.shokusan.or.jp/flexhaccp/intro-3/natto/ (accessed 2022-08-07).

(45) The Role of HACCP in Sanitation. In *Principles of Food Sanitation*; Food Science Text Series; Springer International Publishing: Cham, 2018; pp 109–124. https://doi.org/10.1007/978-3-319-67166-6.

(46) Watanabe, S. *納豆入門 [Introduction to Natto]*; 日本食糧新聞社 [Nihon shokuryō shinbunsha]: Tokyo, 2020.

(47) Watanabe, S. *納豆: 原料大豆の選び方から販売戦略まで [Natto: From how to select raw soybeans to sales strategies]*; 農山漁村文化協会 [Nō-san-gyoson Bunka Kyōkai]: Tokyo, 2002.

(48) 納豆業界の現状と将来 [Current and future of natto industry]. In *納豆入門 [Introduction to Natto]*; 日本食糧新聞社 [Nihon shokuryō shinbunsha]: Tokyo, 2020; pp 145–158.

(49) Japan Natto Cooperative Society Federation. *Research Data*. http://natto.or.jp/reseach/index.html (accessed 2022-12-11).

(50) Kliebenstein, D. J.; Kroymann, J.; Mitchell-Olds, T. The Glucosinolate–Myrosinase System in an Ecological and Evolutionary Context. *Current Opinion in Plant Biology* **2005**, *8* (3), 264–271. https://doi.org/10.1016/j.pbi.2005.03.002.

(51) Tengu Natto. *Dried Natto (Hoshi Natto)*. http://www.tengunatto.com/g-hoshi.html (accessed 2022-12-18).

(52) *Dry Natto Lineup of Overseas Limited | Fried Natto*. 納豆菌が活きている減圧フライ工法のどらいなっとうはフォルッツァ. http://www.dry-natto.com/en/products/ (accessed 2022-12-31).

(53) *Product Information - Natto Snack "Parintou" |*. https://www.sitogi.jp/item/parintou.htm (accessed 2023-01-01).

(54) *Others | Asahimatsu Foods*. https://www.asahimatsu.co.jp/products/others/ (accessed 2023-01-01).

(55) [編者] 納豆試験法研究会, 農林水産省食品総合研究所 = Methods of natto research / Society for Study of Natto, National Food Research Institute, Ministry of Agriculture, Forestry and Fisheries. *納豆試験法 [Methods of Natto Reserach]*, Shohan.; 光琳 [Kōrin]: Tōkyō, 1990.

(56) Morten C. Meilgaard; Gail Vance Civille; B. Thomas Carr. *Sensory Evaluation Techniques*, 4th ed.; CRC Press: Boca Raton, Fla., 2007.

(57) Stone, H.; Bleibaum, R. N.; Thomas, H. A. *Sensory Evaluation Practices*, Fifth edition.; Elsevier/Academic Press: London, United Kingdom, 2021.

(58) Japan Natto Cooperative Society Federation. *Natto Kanpyokai*. http://natto.or.jp/kanpyou/index.html (accessed 2022-12-13).

(59) Japan Natto Cooperative Society Federation. *Natto Shinuchi*. http://natto.or.jp/shinuchi/ (accessed 2022-12-13).

(60) *Food Analysis*; Nielsen, S. S., Ed.; Food Science Text Series; Springer International Publishing: Cham, 2017. https://doi.org/10.1007/978-3-319-45776-5.

(61) 粘質物分析法 [Viscous Material Analysis Method]. In *納豆の科学 [Advanced Science on Natto]*; Kiuchi, K., Nagai, T., Kimura, K., Eds.; 建帛社 [Kenpakusha]: Tokyo, 2008; pp 43–52.

(62) 製品開発事例 [Product Development Case Studies]. In *納豆の研究法 [Investigative Approach to Natto]*; Kan, K., Nagai, T., Kimura, K., Kodaka, K., Muramatsu, K., Watanabe, S., Eds.; 恒星社厚生閣 [Kōseishakō seikaku]: Tokyo, 2010; pp 189–199.

(63) Hayes, D. P. The Growing Inaccessibility of Science. *Nature* **1992**, *356* (6372), 739–740. https://doi.org/10.1038/356739a0.

(64) Eichenberger, P. Genomics and Cellular Biology of Endospore Formation. In *Bacillus: Cellular and Molecular Biology (Third edition)*; Caister Academic Press, 2017; pp 333–366. https://doi.org/10.21775/9781910190579-11.

(65) Earl, A. M.; Losick, R.; Kolter, R. Ecology and Genomics of Bacillus Subtilis. *Trends in Microbiology* **2008**, *16* (6), 269–275. https://doi.org/10.1016/j.tim.2008.03.004.

(66) Nakano, M. M.; Zuber, P. Anaerobiosis. In *Bacillus subtilis and Its Closest Relatives*; John Wiley & Sons, Ltd, 2001; pp 393–404. https://doi.org/10.1128/9781555817992.ch28.

(67) Ivanova, E. P.; Vysotskii, M. V.; Svetashev, V. I.; Nedashkovskaya, O. I.; Gorshkova, N. M.; Mikhailov, V. V.; Yumoto, N.; Shigeri, Y.; Taguchi, T.; Yoshikawa, S. Characterization of Bacillus Strains of Marine Origin. *Int Microbiol* **1999**, *2* (4), 267–271.

(68) Soule, M. H. Identity of Bacillus Subtilis, Cohn 1872. *The Journal of Infectious Diseases* **1932**, *51* (2), 191–215.

(69) Smith, N. R.; Gordon, R. E.; Clark, F. E. *Aerobic Mesophilic Sporeforming Bacteria*; U.S. Department of Agriculture, 1946.

(70) Buchanan, R. E. *Bergey's Manual of Determinative Bacteriology*, 6th ed.; Williams & Wilkins, 1949.

(71) *Bergey's Manual of Determinative Bacteriology*, 8th ed.; Bergey, D. H., Buchanan, R. E., Gibbons, N. E., American Society for Microbiology, Eds.; Williams & Wilkins Co: Baltimore, 1974.

(72) Rosselló-Mora, R.; Amann, R. The Species Concept for Prokaryotes. *FEMS Microbiology Reviews* **2001**, *25* (1), 39–67. https://doi.org/10.1111/j.1574-6976.2001.tb00571.x.

(73) Logan, N. A.; Vos, P. D. Bacillus. In *Bergey's Manual of Systematics of Archaea and Bacteria*; John Wiley & Sons, Ltd, 2015; pp 1–163. https://doi.org/10.1002/9781118960608.gbm00530.

(74) Foster, S. J.; Popham, D. L. Structure and Synthesis of Cell Wall, Spore Cortex, Teichoic Acids, S-Layers, and Capsules. In *Bacillus subtilis and Its Closest Relatives*; John Wiley & Sons, Ltd, 2001; pp 21–41. https://doi.org/10.1128/9781555817992.ch4.

(75) Dijl, J. M. van; Dreisbach, A.; Skwark, M. J.; Sibbald, M. J. J. B.; Tjalsma, H.; Zweers, J. C.; Buist, G. Ins and Outs of the Bacillus Subtilis Membrane Proteome. In *Bacillus: Cellular and Molecular Biology (Third edition)*; Caister Academic Press, 2017; pp 263–294. https://doi.org/10.21775/9781910190579-09.

(76) Graumann, P. L. Chromosome Arrangement and Segregation. In *Bacillus: Cellular and Molecular Biology (Third edition)*; Caister Academic Press, 2017; pp 67–88. https://doi.org/10.21775/9781910190579-03.

(77) Gueiros-Filho, F. Cell Division. In *Bacillus: Cellular and Molecular Biology (Third edition)*; Caister Academic Press, 2017; pp 89–126. https://doi.org/10.21775/9781910190579-04.

(78) Tocheva, E. I.; Ortega, D. R.; Jensen, G. J. Sporulation, Bacterial Cell Envelopes and the Origin of Life. *Nat Rev Microbiol* **2016**, *14* (8), 535–542. https://doi.org/10.1038/nrmicro.2016.85.

(79) Paidhungat, M.; Setlow, P. Spore Germination and Outgrowth. In *Bacillus subtilis and Its Closest Relatives*; John Wiley & Sons, Ltd, 2001; pp 537–548. https://doi.org/10.1128/9781555817992.ch37.

(80) Hazelbauer, G. L. Bacterial Chemotaxis: The Early Years of Molecular Studies. *Annu. Rev. Microbiol.* **2012**, *66* (1), 285–303. https://doi.org/10.1146/annurev-micro-092611-150120.

(81) Aizawa, S.-I.; Zhulin, I. B.; Márquez-Magaña, L.; Ordal, G. W. Chemotaxis and Motility. In *Bacillus subtilis and Its Closest Relatives*; John Wiley & Sons, Ltd, 2001; pp 437–452. https://doi.org/10.1128/9781555817992.ch31.

(82) Hughes, A. C.; Kearns, D. B. Swimming, Swarming and Sliding Motility in Bacillus Subtilis. In *Bacillus: Cellular and Molecular Biology (Third edition)*; Caister Academic Press, 2017; pp 415–438. https://doi.org/10.21775/9781910190579-14.

(83) Maier, B. Competence and Transformation. In *Bacillus: Cellular and Molecular Biology (Third edition)*; Caister Academic Press, 2017; pp 395–414. https://doi.org/10.21775/9781910190579-13.

(84) Okada, M.; Nakamura, Y.; Hayashi, S.; Ozaki, K.; Usami, S. Chemical Structure and Biological Activity of a Quorum Sensing Pheromone from Bacillus Subtilis Subsp. Natto. *Bioorganic & Medicinal Chemistry Letters* **2015**, *25* (19), 4293–4296. https://doi.org/10.1016/j.bmcl.2015.07.083.

(85) Shapiro, J. A. Bacteria as Multicellular Organisms. *Scientific American* **1988**, *258* (6), 82–89.

(86) Aguilar, C.; Vlamakis, H.; Losick, R.; Kolter, R. Thinking about Bacillus Subtilis as a Multicellular Organism. *Current Opinion in Microbiology* **2007**, *10* (6), 638–643. https://doi.org/10.1016/j.mib.2007.09.006.

(87) González-Pastor, J. E. Multicellularity and Social Behaviour in Bacillus Subtilis. In *Bacillus: Cellular and Molecular Biology (Third edition)*; Caister Academic Press, 2017; pp 367–394. https://doi.org/10.21775/9781910190579-12.

(88) Kamada, M.; Hase, S.; Fujii, K.; Miyake, M.; Sato, K.; Kimura, K.; Sakakibara, Y. Whole-Genome Sequencing and Comparative Genome Analysis of Bacillus Subtilis Strains Isolated from Non-Salted Fermented Soybean Foods. *PLOS ONE* **2015**, *10* (10), e0141369. https://doi.org/10.1371/journal.pone.0141369.

(89) Kubo, Y.; Rooney, A. P.; Tsukakoshi, Y.; Nakagawa, R.; Hasegawa, H.; Kimura, K. Phylogenetic Analysis of Bacillus Subtilis Strains Applicable to Natto (Fermented Soybean) Production. *Applied and Environmental Microbiology* **2011**, *77* (18), 6463–6469. https://doi.org/10.1128/AEM.00448-11.

(90) Sumi, H.; Hamada, H.; Tsushima, H.; Mihara, H.; Muraki, H. A Novel Fibrinolytic Enzyme (Nattokinase) in the Vegetable Cheese Natto; a Typical and Popular Soybean

Food in the Japanese Diet. *Experientia* **1987**, *43* (10), 1110–1111. https://doi.org/10.1007/BF01956052.

(91) Danilova, I.; Sharipova, M. The Practical Potential of Bacilli and Their Enzymes for Industrial Production. *Frontiers in Microbiology* **2020**, *11*.

(92) *Natto genome project: Frontpage*. http://natto-genome.org/ (accessed 2022-10-10).

(93) Sharma, C.; Osmolovskiy, A.; Singh, R. Microbial Fibrinolytic Enzymes as Anti-Thrombotics: Production, Characterisation and Prodigious Biopharmaceutical Applications. *Pharmaceutics* **2021**, *13* (11), 1880. https://doi.org/10.3390/pharmaceutics13111880.

(94) *Molecular Genetics of Bacteria*, 4th ed.; Snyder, L., Ed.; ASM Press: Washington, DC, 2013.

(95) Murray, H.; Richardson, T. T.; Noirot-Gros, M.-F.; Polard, P.; Noirot, P. Replication of the Bacillus Subtilis Chromosome. In *Bacillus: Cellular and Molecular Biology (Third edition)*; Caister Academic Press, 2017; pp 1–34. https://doi.org/10.21775/9781910190579-01.

(96) Lewis, P.; Yang, X. The Organisation of Transcription and Translation. In *Bacillus: Cellular and Molecular Biology (Third edition)*; Caister Academic Press, 2017; pp 127–154. https://doi.org/10.21775/9781910190579-05.

(97) De Palmenaer, D.; Siguier, P.; Mahillon, J. IS4 Family Goes Genomic. *BMC Evolutionary Biology* **2008**, *8* (1), 18. https://doi.org/10.1186/1471-2148-8-18.

(98) Nagai, T.; Tran, L. S.; Inatsu, Y.; Itoh, Y. A New IS4 Family Insertion Sequence, IS4Bsu1, Responsible for Genetic Instability of Poly-Gamma-Glutamic Acid Production in Bacillus Subtilis. *J Bacteriol* **2000**, *182* (9), 2387–2392. https://doi.org/10.1128/JB.182.9.2387-2392.2000.

(99) Deutscher, J.; Galinier, A.; Martin-Verstraete, I. Carbohydrate Uptake and Metabolism. In *Bacillus subtilis and Its Closest Relatives*; John Wiley & Sons, Ltd, 2001; pp 129–150. https://doi.org/10.1128/9781555817992.ch11.

(100) Sonenshein, A. L. The Krebs Citric Acid Cycle. In *Bacillus subtilis and Its Closest Relatives*; John Wiley & Sons, Ltd, 2001; pp 151–162. https://doi.org/10.1128/9781555817992.ch12.

(101) von Wachenfeldt, C.; Hederstedt, L. Respiratory Cytochromes, Other Heme Proteins, and Heme Biosynthesis. In *Bacillus subtilis and Its Closest Relatives*; John Wiley & Sons, Ltd, 2001; pp 163–179. https://doi.org/10.1128/9781555817992.ch13.

(102) Luo, Z.; Guo, Y.; Liu, J.; Qiu, H.; Zhao, M.; Zou, W.; Li, S. Microbial Synthesis of Poly-γ-Glutamic Acid: Current Progress, Challenges, and Future Perspectives. *Biotechnology for Biofuels* **2016**, *9* (1), 134. https://doi.org/10.1186/s13068-016-0537-7.

(103) Makoto Ashiuchi. Occurence and Biosynthetic Mechanism of Poly-Gamma-Glutamic Acid. In *Amino-Acid Homopolymers Occurring in Nature*; Hamano, Y., Ed.; Microbiology Monographs; Springer Berlin Heidelberg: Berlin, Heidelberg, 2010; Vol. 15. https://doi.org/10.1007/978-3-642-12453-2.

(104) Keitarou Kimura; Zui Fujimoto. Enzymatic Degradation of Poly-Gamma-Glutamic Acid. In *Amino-Acid Homopolymers Occurring in Nature*; Hamano, Y., Ed.; Microbiology Monographs; Springer Berlin Heidelberg: Berlin, Heidelberg, 2010; Vol. 15. https://doi.org/10.1007/978-3-642-12453-2.

(105) Dogsa, I.; Brloznik, M.; Stopar, D.; Mandic-Mulec, I. Exopolymer Diversity and the Role of Levan in Bacillus Subtilis Biofilms. *PLOS ONE* **2013**, *8* (4), e62044. https://doi.org/10.1371/journal.pone.0062044.

(106) Marvasi, M.; Visscher, P. T.; Casillas Martinez, L. Exopolymeric Substances (EPS) from Bacillus Subtilis : Polymers and Genes Encoding Their Synthesis. *FEMS Microbiology Letters* **2010**, *313* (1), 1–9. https://doi.org/10.1111/j.1574-6968.2010.02085.x.

(107) Kada, S.; Yabusaki, M.; Kaga, T.; Ashida, H.; Yoshida, K. Identification of Two Major Ammonia-Releasing Reactions Involved in Secondary Natto Fermentation. *Bioscience, Biotechnology, and Biochemistry* **2008**, *72* (7), 1869–1876. https://doi.org/10.1271/bbb.80129.

(108) Holt, J. G. *Bergey's Manual of Determinative Bacteriology*; Lippincott Williams & Wilkins, 1994.

(109) Gautam, S. S.; KC, R.; Leong, K. W.; Mac Aogáin, M.; O'Toole, R. F. A Step-by-Step Beginner's Protocol for Whole Genome Sequencing of Human Bacterial Pathogens. *J Biol Methods* **2019**, *6* (1), e110. https://doi.org/10.14440/jbm.2019.276.

(110) Deurenberg, R. H.; Bathoorn, E.; Chlebowicz, M. A.; Couto, N.; Ferdous, M.; García-Cobos, S.; Kooistra-Smid, A. M. D.; Raangs, E. C.; Rosema, S.; Veloo, A. C. M.; Zhou, K.; Friedrich, A. W.; Rossen, J. W. A. Application of next Generation Sequencing in Clinical Microbiology and Infection Prevention. *J Biotechnol* **2017**, *243*, 16–24. https://doi.org/10.1016/j.jbiotec.2016.12.022.

(111) *Whole Genome Sequencing | CDC*. https://www.cdc.gov/pulsenet/pathogens/wgs.html (accessed 2022-11-12).

(112) Vasantha, N.; Freese, E. 1979. The Role of Manganese in Growth and Sporulation of Bacillus Subtilis. *Microbiology 112* (2), 329–336. https://doi.org/10.1099/00221287-112-2-329.

(113) 納豆菌の研究法 [Research Methods of Bacillus natto]. In *納豆の研究法 [Investigative Approach to Natto]*; Kan, K., Nagai, T., Kimura, K., Kodaka, K., Muramatsu, K., Watanabe, S., Eds.; 恒星社厚生閣 [Kōseishakō seikaku]: Tokyo, 2010; pp 1–25.

(114) Jiro Miura; Momojo Yamazaki. *納豆の合理的製造法 [Rational Manufacturing Method of Natto]*; 産業評論社 [Sangyō Hyōronsha]: Tōkyō, 1949.

(115) *Counting Chambers Petroff-Hausser Counter*. http://hausserscientific.com/products/petroff_hausser_counter.html (accessed 2022-11-12).

(116) Altenbuchner, J. Editing of the Bacillus Subtilis Genome by the CRISPR-Cas9 System. *Applied and Environmental Microbiology* **2016**, *82* (17), 5421–5427. https://doi.org/10.1128/AEM.01453-16.

(117) Sachla, A. J.; Alfonso, A. J.; Helmann, J. D. A Simplified Method for CRISPR-Cas9 Engineering of Bacillus Subtilis. *Microbiol Spectr* **2021**, *9* (2), e0075421. https://doi.org/10.1128/Spectrum.00754-21.

(118) Nagai, T.; Itoh, Y. Characterization of a Generalized Transducing Phage of Poly-(Gamma)-Glutamic Acid-Producing Bacillus Subtilis and Its Application for Analysis of Tn917-LTV1 Insertional Mutants Defective in Poly-(Gamma)-Glutamic Acid Production.

Applied and Environmental Microbiology **1997**, *63* (10), 4087–4089. https://doi.org/10.1128/aem.63.10.4087-4089.1997.

(119) Takemura, H. Development of Weak-Smelling Natto. *Journal of the Brewing Society of Japan* **2004**, *99* (11), 794–799. https://doi.org/10.6013/jbrewsocjapan1988.99.794.

(120) Kubo, Y.; Nakagawa, R. Development of Low Sticky Texture Natto "MAMENOKA." *Synthesiology* **2017**, *11* (1), 1–11. https://doi.org/10.5571/synth.11.1_1.

(121) Takahashi, S.; Katsumata, R.; Yoshizawa, K.; Yamaki, M.; Osako, M.; Murakami, M.; Mouri, S.; Sato, M.; Meguro, H.; Kubokura, H.; Hirose, K.; Tanaka, T.; Watanabe, S.; Kiuchi, K. Soft Natto Manufactured with Starter of *Bacillus Subtilis* KFP 843 Isolated from Douchi. *Nippon Shokuhin Kagaku Kogaku Kaishi* **2005**, *52* (10), 454–461. https://doi.org/10.3136/nskkk.52.454.

(122) Slonczewski, J.; Foster, J. W.; Zinser, E. R. *Microbiology: An Evolving Science*, Fifth edition.; W. W. Norton & Company: New York, NY, 2020.

(123) Henkin, T. M.; Peters, J. E. *Snyder and Champness Molecular Genetics of Bacteria*, Fifth edition.; ASM books; John Wiley & Sons, Inc. ; American Society for Microbiology: Hoboken, NJ : Washington, D.C, 2020.

(124) Sonenshein, A. L.; Hoch, J. A.; Losick, R. Bacillus Subtilis: From Cells to Genes and from Genes to Cells. In *Bacillus subtilis and Its Closest Relatives*; John Wiley & Sons, Ltd, 2001; pp 1–5. https://doi.org/10.1128/9781555817992.ch1.

(125) *Bacillus: Cellular and Molecular Biology (Third Edition)*; Peter Graumann, Ed.; Caister Academic Press, 2017. https://doi.org/10.21775/9781910190579.

(126) *納豆の科学 [Advanced Science on Natto]*; Kiuchi, K., Nagai, T., Kimura, K., Eds.; 建帛社 [Kenpakusha]: Tokyo, 2008.

(127) *納豆の研究法 [Investigative Approach to Natto]*; Kan, K., Nagai, T., Kimura, K., Kodaka, K., Muramatsu, K., Watanabe, S., Eds.; 恒星社厚生閣 [Kōseishakō seikaku]: Tokyo, 2010.

(128) Agronomic Characteristics, Production, and Marketing. In *Soybeans: chemistry, technology, and utilization*; Aspen: Gaithersburg, Md., 1997; pp 1–24.

(129) *日本食品標準成分表2020 年版（八訂）：文部科学省 [Standard Tables of Food Composition in Japan 2020 (8th revision)]*. 文部科学省ホームページ. https://www.mext.go.jp/a_menu/syokuhinseibun/mext_01110.html (accessed 2022-11-13).

(130) Guillon, F.; Champ, M. M.-J. Carbohydrate Fractions of Legumes: Uses in Human Nutrition and Potential for Health. *British Journal of Nutrition* **2002**, *88* (S3), 293–306. https://doi.org/10.1079/BJN2002720.

(131) Chemistry and Nutritional Value of Soybean Components. In *Soybeans: chemistry, technology, and utilization*; Aspen: Gaithersburg, Md., 1997; pp 25–113.

(132) Perkins, E. G. Chapter 2 - Composition of Soybeans and Soybean Products. In *Practical Handbook of Soybean Processing and Utilization*; Erickson, D. R., Ed.; AOCS Press, 1995; pp 9–28. https://doi.org/10.1016/B978-0-935315-63-9.50006-1.

(133) *Agronomy Guide for Field Crops - Publication 811*. http://omafra.gov.on.ca/english/crops/pub811/p811toc.html (accessed 2022-08-03).

(134) Orf, J. H. Breeding, Genetics, and Production of Soybeans. In *Soybeans*; Johnson, L. A., White, P. J., Galloway, R., Eds.; AOCS Press, 2008; pp 33–65. https://doi.org/10.1016/B978-1-893997-64-6.50005-6.

(135) Song, J.; Liu, Z.; Hong, H.; Ma, Y.; Tian, L.; Li, X.; Li, Y.-H.; Guan, R.; Guo, Y.; Qiu, L.-J. Identification and Validation of Loci Governing Seed Coat Color by Combining Association Mapping and Bulk Segregation Analysis in Soybean. *PLOS ONE* **2016**, *11* (7), e0159064. https://doi.org/10.1371/journal.pone.0159064.

(136) Government of Canada, C. F. I. A. *Specific work instructions (SWI 142.1.2-6): Soybean seed crop inspection procedures*. https://inspection.canada.ca/plant-health/seeds/seed-inspection-procedures/soybean/eng/1347209394341/1347209477717 (accessed 2023-02-04).

(137) Middelbos, I. S.; Fahey, G. C. Soybean Carbohydrates. In *Soybeans*; Johnson, L. A., White, P. J., Galloway, R., Eds.; AOCS Press, 2008; pp 269–296. https://doi.org/10.1016/B978-1-893997-64-6.50012-3.

(138) Rackis, J. J. Flatulence Caused by Soya and Its Control through Processing. *J Am Oil Chem Soc* **1981**, *58* (3), 503–509. https://doi.org/10.1007/BF02582414.

(139) Morabbi Heravi, K.; Watzlawick, H.; Altenbuchner, J. The MelREDCA Operon Encodes a Utilization System for the Raffinose Family of Oligosaccharides in Bacillus Subtilis. *Journal of Bacteriology* **2019**, *201* (15), e00109-19. https://doi.org/10.1128/JB.00109-19.

(140) Kanno, A.; Takamatsu, H.; Takano, N.; Akimoto, T. Change of Saccharides in Soybeans during Manufacturing of Natto. *Nippon Shokuhin Kogyo Gakkaishi* **1982**, *29* (2), 105–110. https://doi.org/10.3136/nskkk1962.29.2_105.

(141) Murphy, P. A. Soybean Proteins. In *Soybeans*; Johnson, L. A., White, P. J., Galloway, R., Eds.; AOCS Press, 2008; pp 229–267. https://doi.org/10.1016/B978-1-893997-64-6.50011-1.

(142) Harwood, C. R.; Kikuchi, Y. The Ins and Outs of Bacillus Proteases: Activities, Functions and Commercial Significance. *FEMS Microbiology Reviews* **2022**, *46* (1), fuab046. https://doi.org/10.1093/femsre/fuab046.

(143) Kawamura, F.; Doi, R. H. Construction of a Bacillus Subtilis Double Mutant Deficient in Extracellular Alkaline and Neutral Proteases. *Journal of Bacteriology* **1984**, *160* (1), 442–444. https://doi.org/10.1128/jb.160.1.442-444.1984.

(144) Teruo Ohta; Hideo Ebine; Masahiro Nakano; Harukiyo Hieda; Hirokuni Sasaki. Manufacturing New-Type Fermented Soybean Food Product Employing Bacillus Natto - Part 1. Investigation on the Manufacturing on a Laboratory Scale. *Report of the Food Research Institute* **1964**, *18*, 46–52.

(145) The Componential Changes During the Manufacturing Prcoess of Natto (Part 1) -On the Amino Nitrogen ,Ammonia Nitrogen, Carbohydrates and Vitamin B2-.

(146) 明彦飯島; 仁佐々木; 博之若松; 杉夫渡辺; 好美前田. 納豆発酵中の熱量測定 [Calorie measurement during natto fermentation]. 食科工 **1999**, *46* (5), 279–284. https://doi.org/10.3136/nskkk.46.279.

(147) Kimura, K.; Kubo, Y. Flavor Development During Natto Fermentation. *Nippon Shokuhin Kagaku Kogaku Kaishi* **2017**, *64*, 379–384. https://doi.org/10.3136/nskkk.64.379.

(148) Gerde, J. A.; White, P. J. Lipids. In *Soybeans*; Johnson, L. A., White, P. J., Galloway, R., Eds.; AOCS Press, 2008; pp 193–227. https://doi.org/10.1016/B978-1-893997-64-6.50010-X.

(149) Kiuchi, K.; Ohta, T.; Itoh, H.; Takabayahsi, T.; Ebine, H. Studies on Lipids of Natto. *J. Agrc. Food Chem.* **1976**, *24* (2), 404–407.
(150) Lesuisse, E.; Schanck, K.; Colson, C. Purification and Preliminary Characterization of the Extracellular Lipase of Bacillus Subtilis 168, an Extremely Basic PH-Tolerant Enzyme. *European Journal of Biochemistry* **1993**, *216* (1), 155–160. https://doi.org/10.1111/j.1432-1033.1993.tb18127.x.
(151) Wang, T. Minor Constituents and Phytochemicals of Soybeans. In *Soybeans*; Johnson, L. A., White, P. J., Galloway, R., Eds.; AOCS Press, 2008; pp 297–329. https://doi.org/10.1016/B978-1-893997-64-6.50013-5.
(152) Karkle, E. N. L.; Beleia, A. Effect of Soaking and Cooking on Phytate Concentration, Minerals, and Texture of Food-Type Soybeans. *Food Sci. Technol* **2010**, *30*, 1056–1060. https://doi.org/10.1590/S0101-20612010000400034.
(153) Shimizu, M. Purification and Characterization of Phytase from Bacillus Suhtilis (Natto) N–77. *Bioscience, Biotechnology, and Biochemistry* **1992**, *56* (8), 1266–1269. https://doi.org/10.1271/bbb.56.1266.
(154) Feizollahi, E.; Mirmahdi, R. S.; Zoghi, A.; Zijlstra, R. T.; Roopesh, M. S.; Vasanthan, T. Review of the Beneficial and Anti-Nutritional Qualities of Phytic Acid, and Procedures for Removing It from Food Products. *Food Research International* **2021**, *143*, 110284. https://doi.org/10.1016/j.foodres.2021.110284.
(155) Ikeda, R.; Ohta, N.; Watanabe, T. Changes of Isoflavones at Various Stages of Fermentation in Defatted Soybeans. *Nippon Shokuhin Kagaku Kogaku Kaishi* **1995**, *42* (5), 322–327. https://doi.org/10.3136/nskkk.42.322.
(156) Shimakage, A.; Shinbo, M.; Yamada, S.; Ito, H. Changes in Isoflavone Content in Soybeans During the Manufacturing Processes of Tubu-Natto and Hikiwari-Natto. *Nippon Shokuhin Kagaku Kogaku Kaishi* **2006**, *53* (3), 185–188. https://doi.org/10.3136/nskkk.53.185.
(157) Cho, K. M.; Ha, T. J.; Lee, Y. B.; Seo, W. D.; Kim, J. Y.; Ryu, H. W.; Jeong, S. H.; Kang, Y. M.; Lee, J. H. Soluble Phenolics and Antioxidant Properties of Soybean (Glycine Max L.) Cultivars with Varying Seed Coat Colours. *Journal of Functional Foods* **2013**, *5* (3), 1065–1076. https://doi.org/10.1016/j.jff.2013.03.002.
(158) Kubo, Y.; Noguchi, T.; Kimura, K. Storage Temperature and Quality Changes of Natto. *Food Science and Technology Research* **2021**, *27* (3), 497–504. https://doi.org/10.3136/fstr.27.497.
(159) Tanaka, Y.; Tomiyasu, Y. Chemical Components of the White Deposits of Natto (Fermented Soybeans). *Eiyo To Shokuryo* **1973**, *26* (8), 473–478. https://doi.org/10.4327/jsnfs1949.26.473.
(160) Watanabe, A.; Ohtani, T.; Nikkuni, S.; Baba, T.; Ohta, T. The Efficacy of Ultrafiltration Treatment of the Drained Water from Steaming of Soybean in Natto Processing. *Nippon Shokuhin Kogyo Gakkaishi* **1982**, *29* (4), 245–249. https://doi.org/10.3136/nskkk1962.29.4_245.
(161) McGee, H. *Nose Dive: A Field Guide to the World's Smells*; Appetite: Toronto, 2020.
(162) Tanaka, T.; Muramatsu, K.; Kim, H.; Watanabe, T.; Takeyasu, M.; Kanai, Y.; Kiuchi, K. Comparison of Volatile Compounds from Chungkuk-Jang and Itohiki-Natto. *Bioscience,*

Biotechnology, and Biochemistry **1998**, *62* (7), 1440–1444. https://doi.org/10.1271/bbb.62.1440.

(163) *The Good Scents Company - Flavor, Fragrance, Food and Cosmetics Ingredients information*. The Good Scents Company. http://www.thegoodscentscompany.com/ (accessed 2022-12-25).

(164) *Practical Handbook of Soybean Processing and Utilization*; David R. Erickson, Ed.; AOCS Press: Champaign, Illinois, 1995.

(165) Liu, K. *Soybeans: Chemistry, Technology, and Utilization*; Aspen: Gaithersburg, Md., 1997.

(166) Lawrence Alan Johnson; Pamela J. White; Richard Galloway. *Soybeans: Chemistry, Production, Processing, and Utilization*; AOCS Press: Urbana, IL, 2008.

(167) Belitz, H.-D.; Grosch, W.; Schieberle, P. *Food Chemistry*, 3rd rev. ed.; Springer: Berlin, 2004.

(168) *Fennema's Food Chemistry*, Fifth edition.; Damodaran, S., Parkin, K. L., Eds.; CRC Press, Taylor & Francis Group, CRC Press is an imprint of the Taylor & Francis Group, an informa business: Boca Raton, 2017.

(169) Lichtenstein, A. H.; Petersen, K.; Barger, K.; Hansen, K. E.; Anderson, C. A. M.; Baer, D. J.; Lampe, J. W.; Rasmussen, H.; Matthan, N. R. Perspective: Design and Conduct of Human Nutrition Randomized Controlled Trials. *Advances in Nutrition* **2021**, *12* (1), 4–20. https://doi.org/10.1093/advances/nmaa109.

(170) Herreman, L.; Nommensen, P.; Pennings, B.; Laus, M. C. Comprehensive Overview of the Quality of Plant- and Animal-Sourced Proteins Based on the Digestible Indispensable Amino Acid Score. *Food Science & Nutrition* **2020**, *8* (10), 5379–5391. https://doi.org/10.1002/fsn3.1809.

(171) Melina, V.; Craig, W.; Levin, S. Position of the Academy of Nutrition and Dietetics: Vegetarian Diets. *Journal of the Academy of Nutrition and Dietetics* **2016**, *116* (12), 1970–1980. https://doi.org/10.1016/j.jand.2016.09.025.

(172) Ketnawa, S.; Ogawa, Y. In Vitro Protein Digestibility and Biochemical Characteristics of Soaked, Boiled and Fermented Soybeans. *Sci Rep* **2021**, *11* (1), 14257. https://doi.org/10.1038/s41598-021-93451-x.

(173) Nagai, T. Health Benefits of Natto. In *Health Benefits of Fermented Foods and Beverages*; Tamang, J. P., Ed.; CRC Press, Taylor & Francis Group: Boca Raton, 2015.

(174) Kim, I.-S.; Kim, C.-H.; Yang, W.-S. Physiologically Active Molecules and Functional Properties of Soybeans in Human Health—A Current Perspective. *International Journal of Molecular Sciences* **2021**, *22* (8), 4054. https://doi.org/10.3390/ijms22084054.

(175) John W Suttie. Vitamin K. In *Modern nutrition in health and disease*; Wolters Kluwer Health/Lippincott Williams & Wilkins: Philadelphia, 2014; pp 305–316.

(176) Sumi, H. Determination of the Vitamin K (Menaquinone-7) Content in Fermented Soybean Natto and in the Plasma of Natto-Ingesting Subjects. *J. Home Econ. Jpn.* **1999**, *50* (4), 309–312. https://doi.org/10.11428/jhej1987.50.309.

(177) Komori, T. What Is the Function of Osteocalcin? *J Oral Biosci* **2020**, *62* (3), 223–227. https://doi.org/10.1016/j.job.2020.05.004.

(178) Luo, G.; Ducy, P.; McKee, M. D.; Pinero, G. J.; Loyer, E.; Behringer, R. R.; Karsenty, G. Spontaneous Calcification of Arteries and Cartilage in Mice Lacking Matrix GLA Protein. *Nature* **1997**, *386* (6620), 78–81. https://doi.org/10.1038/386078a0.

(179) Hill, M. J. Intestinal Flora and Endogenous Vitamin Synthesis. *European Journal of Cancer Prevention* **1997**, *6* (2), S43.

(180) Weng, Y.; Yao, J.; Sparks, S.; Wang, K. Y. Nattokinase: An Oral Antithrombotic Agent for the Prevention of Cardiovascular Disease. *International Journal of Molecular Sciences* **2017**, *18* (3), 523. https://doi.org/10.3390/ijms18030523.

(181) Nakamura, T.; Yamagata, Y.; Ichishima, E. Nucleotide Sequence of the Subtilisin NAT Gene, AprN, of Bacillus Subtilis (Natto). *Bioscience, Biotechnology, and Biochemistry* **1992**, *56* (11), 1869–1871. https://doi.org/10.1271/bbb.56.1869.

(182) Fujita, M.; Nomura, K.; Hong, K.; Ito, Y.; Asada, A.; Nishimuro, S. Purification and Characterization of a Strong Fibrinolytic Enzyme (Nattokinase) in the Vegetable Cheese Natto, a Popular Soybean Fermented Food in Japan. *Biochem Biophys Res Commun* **1993**, *197* (3), 1340–1347. https://doi.org/10.1006/bbrc.1993.2624.

(183) Fujita, M.; Hong, K.; Ito, Y.; Misawa, S.; Takeuchi, N.; Kariya, K.; Nishimuro, S. Transport of Nattokinase across the Rat Intestinal Tract. *Biol Pharm Bull* **1995**, *18* (9), 1194–1196. https://doi.org/10.1248/bpb.18.1194.

(184) Fujita, M.; Hong, K.; Ito, Y.; Fujii, R.; Kariya, K.; Nishimuro, S. Thrombolytic Effect of Nattokinase on a Chemically Induced Thrombosis Model in Rat. *Biological & Pharmaceutical Bulletin* **1995**, *18* (10), 1387–1391. https://doi.org/10.1248/bpb.18.1387.

(185) Suzuki, Y.; Kondo, K.; Matsumoto, Y.; Zhao, B.-Q.; Otsuguro, K.; Maeda, T.; Tsukamoto, Y.; Urano, T.; Umemura, K. Dietary Supplementation of Fermented Soybean, Natto, Suppresses Intimal Thickening and Modulates the Lysis of Mural Thrombi after Endothelial Injury in Rat Femoral Artery. *Life Sciences* **2003**, *73* (10), 1289–1298. https://doi.org/10.1016/S0024-3205(03)00426-0.

(186) Pais, E.; Alexy, T.; Holsworth, J.; Meiselman, H. J. Effects of Nattokinase, a pro-Fibrinolytic Enzyme, on Red Blood Cell Aggregation and Whole Blood Viscosity. *Clinical Hemorheology and Microcirculation* **2006**, *35* (1–2), 139–142.

(187) Ero, M. P.; Ng, C. M.; Mihailovski, T.; Harvey, N. R.; Lewis, B. H. A Pilot Study on the Serum Pharmacokinetics of Nattokinase in Humans Following a Single, Oral, Daily Dose. *Altern Ther Health Med* **2013**, *19* (3), 16–19.

(188) Kurosawa, Y.; Nirengi, S.; Homma, T.; Esaki, K.; Ohta, M.; Clark, J. F.; Hamaoka, T. A Single-Dose of Oral Nattokinase Potentiates Thrombolysis and Anti-Coagulation Profiles. *Sci Rep* **2015**, *5* (1), 11601. https://doi.org/10.1038/srep11601.

(189) Kim, J. Y.; Gum, S. N.; Paik, J. K.; Lim, H. H.; Kim, K.; Ogasawara, K.; Inoue, K.; Park, S.; Jang, Y.; Lee, J. H. Effects of Nattokinase on Blood Pressure: A Randomized, Controlled Trial. *Hypertens Res* **2008**, *31* (8), 1583–1588. https://doi.org/10.1291/hypres.31.1583.

(190) Jensen, G. S.; Lenninger, M.; Ero, M. P.; Benson, K. F. Consumption of Nattokinase Is Associated with Reduced Blood Pressure and von Willebrand Factor, a Cardiovascular Risk Marker: Results from a Randomized, Double-Blind, Placebo-Controlled,

Multicenter North American Clinical Trial. *Integr Blood Press Control* **2016**, *9*, 95–104. https://doi.org/10.2147/IBPC.S99553.

(191) Hodis, H. N.; Mack, W. J.; Meiselman, H. J.; Kalra, V.; Liebman, H.; Hwang-Levine, J.; Dustin, L.; Kono, N.; Mert, M.; Wenby, R. B.; Huesca, E.; Rochanda, L.; Li, Y.; Yan, M.; St John, J. A.; Whitfield, L. Nattokinase Atherothrombotic Prevention Study: A Randomized Controlled Trial. *Clin Hemorheol Microcirc* **2021**, *78* (4), 339–353. https://doi.org/10.3233/CH-211147.

(192) Wen, H.; Park, K. *Oral Controlled Release Formulation Design and Drug Delivery: Theory to Practice*; Wiley: Hoboken, N.J., 2010.

(193) Dressman, J. B.; Berardi, R. R.; Dermentzoglou, L. C.; Russell, T. L.; Schmaltz, S. P.; Barnett, J. L.; Jarvenpaa, K. M. Upper Gastrointestinal (GI) PH in Young, Healthy Men and Women. *Pharm Res* **1990**, *7* (7), 756–761. https://doi.org/10.1023/a:1015827908309.

(194) Gardner, J. D.; Ciociola, A. A.; Robinson, M. Measurement of Meal-Stimulated Gastric Acid Secretion by in Vivo Gastric Autotitration. *Journal of Applied Physiology* **2002**, *92* (2), 427–434. https://doi.org/10.1152/japplphysiol.00956.2001.

(195) Read, N. W.; Cammack, J.; Edwards, C.; Holgate, A. M.; Cann, P. A.; Brown, C. Is the Transit Time of a Meal through the Small Intestine Related to the Rate at Which It Leaves the Stomach? *Gut* **1982**, *23* (10), 824–828. https://doi.org/10.1136/gut.23.10.824.

(196) Lorkowski, G. Gastrointestinal Absorption and Biological Activities of Serine and Cysteine Proteases of Animal and Plant Origin: Review on Absorption of Serine and Cysteine Proteases. *Int J Physiol Pathophysiol Pharmacol* **2012**, *4* (1), 10–27.

(197) Tanimoto, H.; Mori, M.; Motoki, M.; Torii, K.; Kadowaki, M.; Noguchi, T. Natto Mucilage Containing Poly-γ-Glutamic Acid Increases Soluble Calcium in the Rat Small Intestine. *Bioscience, Biotechnology, and Biochemistry* **2001**, *65* (3), 516–521. https://doi.org/10.1271/bbb.65.516.

(198) Lee, M.-S.; Kang, J.-I.; Kim, H.-S. Effect of γ-PGA (Poly-γ-Glutamic Acid) Supplement on Calcium Absorption and Bone Metabolism in Rats. *Journal of the Korean Society of Food Science and Nutrition* **2006**, *35* (3), 255–261. https://doi.org/10.3746/jkfn.2006.35.3.255.

(199) Terada, A.; Yamamoto, M.; Yoshimura, E. Effect of the Fermented Soybean Product "Natto" on the Composition and Metabolic Activity of the Human Fecal Flora. *Jpn. J. Food Microbiol.* **1999**, *16* (4), 221–230. https://doi.org/10.5803/jsfm.16.221.

(200) Adamberg, K.; Tomson, K.; Talve, T.; Pudova, K.; Puurand, M.; Visnapuu, T.; Alamäe, T.; Adamberg, S. Levan Enhances Associated Growth of Bacteroides, Escherichia, Streptococcus and Faecalibacterium in Fecal Microbiota. *PLOS ONE* **2015**, *10* (12), e0144042. https://doi.org/10.1371/journal.pone.0144042.

(201) Kim, I.-S. Current Perspectives on the Beneficial Effects of Soybean Isoflavones and Their Metabolites for Humans. *Antioxidants (Basel)* **2021**, *10* (7), 1064. https://doi.org/10.3390/antiox10071064.

(202) Reed, K. E.; Camargo, J.; Hamilton-Reeves, J.; Kurzer, M.; Messina, M. Neither Soy nor Isoflavone Intake Affects Male Reproductive Hormones: An Expanded and Updated Meta-Analysis of Clinical Studies. *Reprod Toxicol* **2021**, *100*, 60–67. https://doi.org/10.1016/j.reprotox.2020.12.019.

(203) Nagata, C. Factors to Consider in the Association Between Soy Isoflavone Intake and Breast Cancer Risk. *Journal of Epidemiology* **2010**, *20* (2), 83–89. https://doi.org/10.2188/jea.JE20090181.

(204) Kim, B.; Byun, B. Y.; Mah, J.-H. Biogenic Amine Formation and Bacterial Contribution in Natto Products. *Food Chemistry* **2012**, *135* (3), 2005–2011. https://doi.org/10.1016/j.foodchem.2012.06.091.

(205) Nuñez, M.; del Olmo, A.; Calzada, J. Biogenic Amines. In *Encyclopedia of Food and Health*; Caballero, B., Finglas, P. M., Toldrá, F., Eds.; Academic Press: Oxford, 2016; pp 416–423. https://doi.org/10.1016/B978-0-12-384947-2.00070-2.

(206) Inomata, N.; Osuna, H.; Ikezawa, Z. Late-Onset Anaphylaxis to Bacillus Natto–Fermented Soybeans (Natto). *Journal of Allergy and Clinical Immunology* **2004**, *113* (5), 998–1000. https://doi.org/10.1016/j.jaci.2004.02.037.

(207) Nagata, C.; Wada, K.; Tamura, T.; Konishi, K.; Goto, Y.; Koda, S.; Kawachi, T.; Tsuji, M.; Nakamura, K. Dietary Soy and Natto Intake and Cardiovascular Disease Mortality in Japanese Adults: The Takayama Study. *The American Journal of Clinical Nutrition* **2017**, *105* (2), 426–431. https://doi.org/10.3945/ajcn.116.137281.

(208) Ross, A. C. *Modern Nutrition in Health and Disease*, 11th ed.; Wolters Kluwer Health/Lippincott Williams & Wilkins: Philadelphia, 2014.

(209) Hosoi, T.; Kiuchi, K. Natto: A Soybean Food Made by Fermenting Cooked Soybeans with Bacillus Subtilis (Natto). In *Handbook of Fermented Functional Foods*; Farnworth, E. R., Ed.; Functional foods & nutraceuticals series; CRC Press: Boca Raton, 2008.

(210) *History of Fermented Black Soybeans (165 B.C. to 2011) - SoyInfo Center.* https://www.soyinfocenter.com/books/150 (accessed 2022-11-26).

(211) Zhang, Y.; Zeng, T.; Wang, H.; Song, J.; Suo, H. Correlation between the Quality and Microbial Community of Natural-Type and Artificial-Type Yongchuan Douchi. *LWT* **2021**, *140*, 110788. https://doi.org/10.1016/j.lwt.2020.110788.

(212) Peng, Y.; Huang, Q.; Zhang, R.; Zhang, Y. Purification and Characterization of a Fibrinolytic Enzyme Produced by Bacillus Amyloliquefaciens DC-4 Screened from Douchi, a Traditional Chinese Soybean Food. *Comparative Biochemistry and Physiology Part B: Biochemistry and Molecular Biology* **2003**, *134* (1), 45–52. https://doi.org/10.1016/S1096-4959(02)00183-5.

(213) Hu, Y.; Yu, D.; Wang, Z.; Hou, J.; Tyagi, R.; Liang, Y.; Hu, Y. Purification and Characterization of a Novel, Highly Potent Fibrinolytic Enzyme from Bacillus Subtilis DC27 Screened from Douchi, a Traditional Chinese Fermented Soybean Food. *Sci Rep* **2019**, *9* (1), 9235. https://doi.org/10.1038/s41598-019-45686-y.

(214) Kon, M.; Ito, H. Studies on "Hama-Natto" (Part 2). *Journal of Home Economics of Japan* **1974**, *25* (1), 21–26. https://doi.org/10.11428/jhej1951.25.21.

(215) 寛記伊藤. Hama-Natto 浜納豆. 醸協 **1976**, *71* (3), 173–176. https://doi.org/10.6013/jbrewsocjapan1915.71.173.

(216) Chukeatirote, E. Thua Nao: Thai Fermented Soybean. *Journal of Ethnic Foods* **2015**, *2*. https://doi.org/10.1016/j.jef.2015.08.004.

(217) Toshirou Nagai; Tamang, J. P. Fermented Legumes: Soybean and Non-Soybean Products. In *Fermented Foods and Beverages of the World*; Kailasapathy, K., Ed.; CRC Press: Boca Raton, 2010; pp 191–224. https://doi.org/10.1201/EBK1420094954.

(218) Tamang, J. P. *Himalayan Fermented Foods: Microbiology, Nutrition, and Ethnic Values*; Taylor and Francis Group: New York, 2010.

(219) Mine, Y.; Kwan Wong, A. H.; Jiang, B. Fibrinolytic Enzymes in Asian Traditional Fermented Foods. *Food Research International* **2005**, *38* (3), 243–250. https://doi.org/10.1016/j.foodres.2004.04.008.

(220) Tamang, J. P. Naturally Fermented Ethnic Soybean Foods of India. *Journal of Ethnic Foods* **2015**, *2* (1), 8–17. https://doi.org/10.1016/j.jef.2015.02.003.

(221) *Health Benefits of Fermented Foods and Beverages*; Tamang, J. P., Ed.; CRC Press, Taylor & Francis Group: Boca Raton, 2015.

(222) *Fermented Foods and Beverages of the World*; Tamang, J. P., Kailasapathy, K., Eds.; CRC Press: Boca Raton, 2010. https://doi.org/10.1201/EBK1420094954.

Index

Q

R

S

About the Author

Keith Wong is a chemical engineer based in Ontario, Canada. He enjoys microbiology and bioprocess engineering, in food and in water, wastewater, and solid waste treatment.

Blog: controlledmold.com
Twitter: @Keeifa

Made in the USA
Monee, IL
11 November 2026

ddb213ec-2a2e-46be-8779-156c062243bdR01